Palgrave Studies in Sound

Series Editor
Mark Grimshaw-Aagaard
Musik
Aalborg University
Aalborg, Denmark

Palgrave Studies in Sound is an interdisciplinary series devoted to the topic of sound with each volume framing and focusing on sound as it is conceptualized in a specific context or field. In its broad reach, Studies in Sound aims to illuminate not only the diversity and complexity of our understanding and experience of sound but also the myriad ways in which sound is conceptualized and utilized in diverse domains. The series is edited by Mark Grimshaw-Aagaard, The Obel Professor of Music at Aalborg University, and is curated by members of the university's Music and Sound Knowledge Group.

Editorial Board
Mark Grimshaw-Aagaard (series editor)
Martin Knakkergaard
Mads Walther-Hansen
Kristine Ringsager

Editorial Committee
Michael Bull
Barry Truax
Trevor Cox
Karen Collins

More information about this series at
http://www.palgrave.com/gp/series/15081

Mickey Vallee

Sounding Bodies Sounding Worlds

An Exploration of Embodiments in Sound

Mickey Vallee
Centre for Interdisciplinary Studies
Athabasca University
Athabasca, AB, Canada

Palgrave Studies in Sound
ISBN 978-981-32-9326-7   ISBN 978-981-32-9327-4   (eBook)
https://doi.org/10.1007/978-981-32-9327-4

This Palgrave Macmillan imprint is published by the registered company Springer Nature Singapore Pte Ltd.
The registered company address is: 152 Beach Road, #21-01/04 Gateway East, Singapore 189721, Singapore

For Winter and for Wren
My favourite season and favourite bird

Acknowledgements

This book was written on Treaty 6 (the ancestral and traditional territory of the Cree, Dene, Blackfoot, Saulteaux, Nakota Sioux, as well as the Métis), Treaty 7 (the ancestral and traditional territory of the Blackfoot Confederacy: Kainai, Piikani and Siksika as well as the Tsuu T'ina First Nation and Stoney Nakoda First Nation), and Treaty 8 (the ancestral and traditional territory of the Cree, Dene, as well as the Métis) territories.

Much of the empirical research for this book was undertaken in field sites from 2016 to 2018 around Alberta, Canada, at Waterton Lakes National Park, Wood Buffalo National Park, and the Frank Slide Interpretive Centre, as well as the Bayne Lab at the University of Alberta. I am particularly grateful to the time provided by Erin Bayne and Elly Knight at the University of Alberta, and Helena Mahoney and Barb Johnston at Waterton, for introducing me to the wonderful world of bioacoustics. The more theoretical endeavours of this book were made possible through conversations facilitated by the Posthumanism Research Institute, especially by Christine Daigle at Brock University and Russel Kilbourn at Wilfrid Laurier University. The preliminary stage of this research was facilitated by Jan Jagodzinsky's generous invitation to speak to the issue of sound and environmental uncertainty at the University of Alberta in 2015. My graduate students, Marie-Josée Beaulieu, David Stewart, and Priscilla McGreer, provided many useful comments on interdisciplinary research, as well. I also benefited greatly from the many conversations with my former doctoral supervisor and now research collaborator, Rob Shields, whose many conversations still serve as a guiding light for much of my work. I am also grateful to Michael Gallagher, who invited me to speak at Manchester Metropolitan University on the topic of voice and sound, and who also commented on sections of the book with generosity and openness. Ada Jaarsma and Simon Dawes also commented on sections of the work that were of enormous help.

I would like to extend a special thanks to Mark Grimshaw-Aagaard for believing in this work, and to Jonathan Sterne for his preliminary guidance and comments on the project. Senior Commissioning Editor Josh Pitt demonstrated patience and goodwill while I was writing and editing this book. DL Wybert's editorial comments taught me much about the craft of writing.

Countless conversations and discussions go into the making of research, and there are too many to count from over the years. Other nodes in the network of the support include Jason Wallin, Sourayan Mookerjea, Natalie Loveless, Harvey Krahn, Paul Théberge, Jonathan Sterne, Sha Xin Wei, Will Straw, Trevor Pinch, Stefan Helmreich, Julie Laplante, Susan McDaniel, Jennifer Gabrys, Ondine Park, Cary Wolfe, JH, Andriko Lozowy, Carolina Cambre, Francesca Mackenney, W. Tecumseh Fitch, Jody Berland, John Shepherd, Mike Gismondi, Paul Kellogg, Wendell Kesner, Max Ritts, and many others whose mistaken absence from this list should not imply ingratitude. I have benefited greatly from the many people I met and talked to at the annual meetings for the Society for the Social Studies of Science, Society for Literature, Science and the Arts, Society for Existential and Phenomenological Theories of Culture, Canadian Sociological Association, Canadian Communication Association, and American Sociological Association.

Much of this book is new. But bits and earlier drafts of the chapters have appeared in such venues as *Theory, Culture & Society, Body &*

Society, Parallax, Space & Culture, The Oxford Handbook of Sound and the Imagination (Vol. 2), and *Interrogating the Anthropocene* (Palgrave, 2018).

The research for this book was undertaken, in part, thanks to funding from the Canada Research Chairs program. It was also supported generously by an Insight Development Grant from the Social Science and Humanities Research Council of Canada, as well as a John R Evans Leaders Funds award from the Canada Foundation for Innovation.

Contents

1

Introduction: Sounding

How does a world sound? While diverse answers to this question emerge depending on perspective, I adopt a media theory approach here that frames an answer in the technologies that open onto objective discoveries about sounding worlds, along with the new intimacies between researchers and those fields. This book deploys some of those technologies in disparate ways to make its point—through laryngoscopes and vocal folds, databases and biodiversity, orangutans and grunts, nets and birds, rockslides and tourism. In short, the trans-species and transtechnological futures of sound become a method for understanding how we might build relations and communities that are trans-species, transtechnological, and transdisciplinary. Sound does not arrive without the infrastructure that anticipates its arrival, and as much as sound arrives to give us information about the system from which it emerges, it remains an integral part of how that system holds together. We cannot shut our bodies off to the vibrations that surround us. But we can turn our attention towards the direction of these vibrations and make sense of the material systems with which they resonate; they resonate at once in the technological objects of measure as well as in the objective reality that determines our interests.

© The Author(s) 2020

M. Vallee, *Sounding Bodies Sounding Worlds*, Palgrave Studies in Sound,
https://doi.org/10.1007/978-981-32-9327-4_1

What we do *with* sound is much less straightforward than what we do with images. Where images produce bodies that have defined edges and boundaries, sound conjoins bodies and bleeds much more readily between them—while images are fixed, sounds flow. Discovery in sound comes with a paradoxical abstract certainty: When a bird is identified, when a rattle in the breath is detected, when the grunt of an orangutan is discovered, when the algorithm detects a future pattern of biodiversity loss through the distribution of soundscapes—such a detection through sound produces palpable imaginaries of what we connect with. This book draws upon such examples to convey how a world sounds.

This book resonates with Jussi Parikka's (2010) idea of an "ethological media theory," which he claims "leads us to evaluate bodies not according to their innate, morphological essences but as expressions of certain movements, sensations, and interactions with their environments" (p. xxv). Intent on the generative capacity of technology to "sound out" the cultural meanings of research across the natural and social sciences, arts and humanities, this book offers a robust cultural theory of sounding, so-called by some sound studies scholars. Sounding encompasses the infrastructural labour that goes into the production of sound, and the mental attitude towards sound as a *method* for understanding acoustic embodiments in a wide range of circumstances—from human voices (Chapter 2), transspecies voices (Chapter 3), transacoustic communities (Chapter 4), data sound (Chapter 5), and geoacoustic vibrations (Chapter 6). Indeed, the range of technologies that accompany each of these contexts shows us that an acoustic embodiment is never only an abstract, but rather, also *concrete* and *situated.* The technologies pursued here make for a broad multisensory, multiscalar, perspective and pushes on the threshold of what sounds, a propagation between bodies, technology, science, and culture.

However, this book is also concerned with how listening has become compressed and displaced, assigned to nonhuman actors and technical assemblages, peripheral to the experiences of humans, but directly implicating them nonetheless; the infrastructure of audibility, a term that will be used throughout this book, now must include, along with human listeners, animal listeners and machine listeners, plant listeners and mountain listeners. If listening is usually placed at the centre of experience, here listening has become a computational category, like other modalities.

But in being so, listening has also adopted new ethical dimensions regarding conservation and ethical trans-species relations. Sounding here is thus related to the practices around and through which sound is produced by both bodies and interrelationality, via the material production that underlies sounding—what I term infrastructures of audibility and imaginary organs.

Sounding: Infrastructures of Audibility and Imaginary Organs

The central theory of the book is one of *sounding*, and it is a theory which is bound by two concepts that traverse a range of case studies from the human body, to bioacoustics, to the vibrations of place. The theory of sounding points to two concepts that will be explicated throughout the book: infrastructures of audibility and imaginary organs. These will be explicated shortly, but first, a word on where my use of the term sounding comes from. Biological anthropologist and cultural theorist Stefan Helmreich's (2015) use of the term *sounding* is intended to describe how we bring the unknowable into the limits of the knowable—how to bring things from outside our field of vision into view through sound, how to generate worlds. Such a term is generalizable enough that it accounts for how new technologies bring together new communities in vibration with one another, because they touch by approximation *in vibration*; these may be interhuman communities (such as Deaf and Hearing communities) and interspecies communities (researcher and animal), but also extra-human communities (plant and sun rays).

Helmreich (2015) writes that sounding is an imaginative device that intends to capture the entanglements of scientific research and cultural formations: The basis of this formation is the propagation and persistence of life. Sounding is thus situated, concrete—an actualization. As such, sounding is radically empirical and resists overly theorizing ventures. This means that sounding is a method that interrogates objects in the world and is especially useful for gaining closeness and proximity to them. In particular, Helmreich writes that sound is more than a colloquial experience in that it resonates. As such, the meaning links back to the etymological roots of the English *sund,* meaning sea, implying something infused with deep sounds. Thus, as an experiential concept, sounding helps us understand the *experience* of sound as opposed to sound as an *event or a phenomenon* itself. Instead, it invokes the collective phenomena that go into the labour and the process of sounding; as we will see in many of the examples, sounding involves material and extramaterial labour, some of it quite visual, much of it markedly haptic.

Sounding means getting caught up in the excitement of multiple meanings in a sound, and of perceiving the meanings of sound in non-sonic or trans-sonic things (Chapter 4 addresses this in terms of a *transacoustic community*). It also takes into account the ambivalences of the abstract and the material; examples and field trips are as much abstractions as they are actualizations. Helmreich (2015) argues that sounding is thus the most appropriate word for approximating what is not yet known. He defines it, within empirical research, as

> fathoming, resounding, uttering, being heard, conveying impressions, suggesting analogies, repeating, and echoing….[It is] an appropriate idiom for investigating that which is not yet fully known, that which people discover only through a kind of auditing that can change the very substance to which it listens, that can create new echoes, new reverberations. (2015, p. 187)

The technologies through which these resonances are realized create the potential for a more intimate connection between individual and collective identities. They also give more direct and accurate access to inner mechanisms of the social and natural worlds. Sounding is a mode of practical theorizing, a way of approaching parts of the empirical world that otherwise elude us, and of participating in sound gathering and analysis. But it is also a way of *attuning to the technological and infrastructural relations that underlie and open onto experience.*

While Helmreich's (2015) determination is correct, I find his terminological assertiveness on sounding ambiguous. Does sounding have anything to do with the infrastructures of sound, beyond metaphoric slippages? Here, I do maintain Helmreich's idea of sounding—but specifically to open it in a new direction, with all the generosity that his term bestows. Helmreich compels us to take seriously the alignment of sound with all things affective, atmospheric, and virtual, as is the fashion lately. In this book, sound will instead refer to its own assemblage of technologies through which sounding sounds. This frames sound *as an event*, and all those who contribute towards sound as participants in "doubleness," according to current digital studies terminology: such mediation opens onto new worlds while contributing to the shape of those worlds (see Hansen, 2015, pp. 6–8). This mediation generates new knowledge while producing new positions of power, where observations lead to improved knowledge, cohering into new sites of mediation. Sounding thus reflects a binary term, a body facing two directions that co-produce one another—*infrastructures of audibility* and *imaginary organs*:

- *Infrastructures of audibility.* This is not the same as an audible infrastructure, which is far more often the focus for those with an interest in sonification, and would imply a route of sonification that can be listened to by (mostly) human ears. Infrastructures of audibility refers to the conduits and occupational labour that facilitate the flow of sound and that undergird knowledge. Infrastructures of audibility do not describe sound so much as the material systems through which the production of sound is made possible. It refers at once to the technologies of sound production, but moreover, to their networks and their ecologies. In some cases, it may seem as though an infrastructure of audibility is not about sound at all, but rather about the underlying vibrational atmospheres that those infrastructures produce. The making of sound is crucial to the concept of an infrastructure of audibility. I choose audibility because this word implies an ability to hear, whether that is a human or a nonhuman species, or a machine or an algorithm. Audibility expands sound from its human-centricity, but not at the expense of human experience altogether. The question of *to whom* or *to what* is sound audible is one which forces us to expand our focus of sound. Infrastructure, meanwhile, has remained. A point of interest in the social sciences as the interactions between humans and the hardware upon which their experience is spatialized; or, as anthropologist Brian Larkin (2013) has written, "matter that enable the movement of other matter" (p. 329).

- *Imaginary organs*: The infrastructure of audibility produces imaginary organs; what is *captured as datum* is the invariant in a form that is recognizable, like how we know the hoarseness of a voice signals fatigue. It is our recognition that allows us to speak of it, the sound, as an imaginary organ. I choose *organ* in place of object, thing, or instrument for a few reasons. First, an *object* implies a divide between subject and object. A *thing*, meanwhile, implies an inertia, what it *will do*, what its *potential is*, as if its agency lies in its matter and not so much the matter of the humans and nonhumans who interact with it. This does not capture the liveliness of what bodies do with things, or what bodies do when they perceive themselves, through themselves, as things. An *instrument* may be acceptable but reads too closely to an instrumental approach or an instrumentalization. Thus, organ is preferred: it is playfully musical, denoting an instrument, and a quasi-object with the potential of a thing; but *organ* also denotes *connection, life,* and *breathing through.* If sounding is a theory for a relational vibration, the *imaginary organ* is the infrastructure of audibility for sounding. *Imaginary,* meanwhile, refers to something social and productive, yet something produced through the sociotechnical infrastructure of audibility that brings forth the imaginary organ.

Finally, to address the broader ethical stakes underlying this book: The ethics in this book are concerned with openness, and openings between bodies; infrastructures generally serve to determine openings and openness of systems. Openings are ethical insofar as an opening is not something that we think, but something that is lived in and through a body in its relations to other bodies. All bodies open, and all openings refer to the oscillatory movement of a body, and all bodies vibrate in response to the vibration of other bodies, whether those bodies are inside another body (like a larynx) or external to it (like a rockslide). And unlike sound, which we hear as the product of invariance, vibration is lived through unconsciously: it is mainly and really only thought about when it is something new and disruptive, something which changes the way we live. It is something we measure, feel, and live out, not something that we think. But we also do not *live* in vibration, but rather live through the openings that vibrations afford.

As an element of an epistemic field, we can shift focus towards what sounding offers us as new knowledge in the milieu of an infrastructural audibility. As discussed in Chapter 4, for instance, researchers mount onto nighthawks miniature microphone backpacks cross-correlated with back-pack GPS devices so that they can determine the specificity of the birds' behaviours as they fly and forage. But the infrastructural labour behind installing them, the value that scientists place on sound, on resonance and voice and embodiment, on the proximity of the body, are the domains of special interest—the recording apparatus in this particular case con-sists of technologies that, on their own, have little to do with sound, such as nets and aluminium tubes, pickup trucks and fatbikes, fire and sand. But together, they are all about the sound in their precipitation and anticipation of the sound event—the infrastructure of audibility is a set of openings towards the production of an imaginary organ, a nighthawk sound in flight. The infrastructure of audibility of Chapter 5 is quite a different matter in terms of its significance to contemporary scientific research (why and how it matters in the research agendas of scientists). This chapter considers why we might be compelled to employ a radically computational approach to sound, which takes into account the vibra-tional resonances that anticipate the absence of human presence; within this perspective, all elements of sounding are intended to eradicate the human perspective, to make sounding and sensation radically computa-tional, and to embed that computation into environments in which it matters very little whether humans are present to hear sounds or not. What is infrastructurally audible may bear very little relevance to whether a human can hear it.

In terms of imaginary organs, this refers to the object of sound which is the vibration between entities in motion, and refers explicitly to a mate-rialist ground in sounding. In sound studies, we have adapted to the well-worn argument that new technologies in sounding prescribe, but do not necessarily determine, new scales and patterns of listening, which in turn contribute towards sounding in new and creative ways. This is a perspective shared among sound scholars regarding other technologies, commonly directed towards the relationship between object and practice. For instance, Jody Berland (2009) claimed that the leisurely domestic practices of the nineteenth-century middle-class home were organized around the family piano; Mark Katz (2010) argued that the phonograph introduced a "phonograph effect" that set in motion changes to and refinement of private and public listening. The phonograph effect is extended to "any observable manifestation of recording's influence" (Katz, 2010, p. 2). However, this perspective is challenged by Jonathan Sterne's (2003) foundational work in sound studies, which argues that sound reproduction technology was precipitated by variation of practices and ideas *about* sound reproduction in scientific and medical experiments. Such practices also precipitated the individuated collectivity of listening, where listening habits became individually attuned to yield unique perspectives on a common object, such as multiple stethoscope tubes for a single patient or listening tubes for a single phonograph recording. Media objects, Sterne has established, thus determined patterns of action and interaction, but they were designed with other pre-existent patterns well in place, *organized* by *imagination*.

One of the most interesting moves in sound studies has been towards the epistemic aspects of sound, how we know in and through sound. In her recent book, *Sonic Skills*, Karin Bijsterveld (2019) examines the sonic dimensions around knowledge-making, particularly in relation to the methodological scope of researchers in the hard sciences. Her interest is "in sound and listening as a way of acquiring knowledge about human bodies, animals, machines, or other research objects, and thus in sound and listening as a means rather than an object of research" (p. 4). She writes that listening constitutes an embodied and inward sense and sensation of knowledge-making in the sciences, which involves an intimate relation between the bodies of the researcher and the researched: a physical proximity is conjoined by the touch of vibration. For her, listening is a social practice embedded in the developments of science and technology; our specific listening practices (especially those as they pertain to the sciences of the inner body) are tuned to extract particular information about changes to the inner state of the self.

Together, the concepts of the infrastructures of audibility and the imaginary organ constitute a media theory of *sounding*. The two concepts underlying the theory of sounding point to new directions in the sonic dimensions of science and technology studies (STS). While STS scholars have become steadily interested in the usefulness of sound and sounding as a method instead of as an object, it is understandable that throughout the twentieth and the twenty-first century sound has become entwined in a knot of theories about mediations, transductions, and transformations. Indeed, as Trevor Pinch and Karin Bijsterveld (2012) have written in their *Oxford Handbook of Sound Studies*, sound is imbricated in mediations that are at once schizophonic, cut off from their source, while generating new intimacies between listeners and devices. Further, as Barry Truax (2001) has noted in his art/research practice of acoustic ecology, listening is an active form of giving attention to something, in that it makes an effort to open onto the information disclosed by another.

I argue that sounding is entangled in complex technical networks. As such, the book intensifies the ethics of sounding to include new openings provided by infrastructures of audibility. These mediations relate, in particular, to the search for shared ancestry in primate communication (Chapter 3), the ecological manifestation of sound and the animal body (Chapter 4), the vibrations of the earth in landslide disasters (Chapter 6), and the finitude of vibration as it can be sliced into by emerging sonic technologies (concluding chapter). In some cases, sounding has become so radicalized that human reception has been displaced onto emerging sound technologies that are programmed to render sound audible only to computation (Chapter 5). In particular, I engage with the ethics of embodiment, ecology, and geological mediations to propose a new and modest concept in the material systems of sounding.

My aim is to capture the collective activity infusing the production and openness of sound as it is deployed through scientific research, and to unpack that utilization to shed new light on cultural theories of embodiment and materiality. Indeed, this book proposes the controversial view that listening has become something of a democratic norm in the sciences. This state arises because listening is no longer a directly embodied phenomenon but something that can be programmed and is programmable into new emerging technologies, which are intended to extract information from deep within bodies and from habitats within which bodies interact and make sense of things. In the context of deepening concerns over the future of the environment and of the continued relevance of biodiversity in a world that is increasingly indifferent or even hostile to difference, such a project is pertinent to find new ethical relations between people and their earthly cohabitants.

Sound persists on the threshold of visibility, but I wish to explore throughout this book those thresholds we call "sound." What are the convergences of the heard and the unheard, the sound and the unsound, and how do we make sense of these convergences? Taken from the perspective of a media materialist ethics, sounding is propagated through current scientific practices in listening and digital mediations. Building on (and contributing to) current conversations and debates in media theory about the notion of the elemental and the atmospheric, the imperceptible and the affective, the virtual and the actual, I propose that the ethics of sounding requires at once embodied empirics, technical entanglements, and a connection to virtual gatherings.

The book makes a contribution to media theory by invoking the unprecedented opportunity to reframe our conception of agency and environmental communities according to the new complex infrastructural media that characterize the contemporary mediascape. I theorize such infrastructural mediation along several angles, from environmental governmentality, to intimacy in science and technology, to media archaeology. I build upon mediation by surveying the ontological inventiveness of sounding practices in the sciences and arts, connecting their infrastructures of audibility, and in the process reconfiguring how we can understand the production of scientific knowledge through imaginary organs.

Sound as an Object of Study

Sounding is ethical and relational. Indeed, scientific studies in which sound is used to gather data are not likely to privatize (or share) a listening experience, but rather to project their findings onto public digital archives for the purposes of community engagement and knowledge mobilization. Joeri Bruyninckx (2018), for example, points to the growing collaborations between "citizen scientists" and the Cornell Library of Natural Sounds, which is in need of the public's hobby recordings to expand its growing catalogue of bird songs. The collaborative and empirical use of sound reinforces Julian Henriques's (2011) thesis that "sounding" binds together everyone and everything involved in the production of sound. In the context of my own study, that includes human (researchers), nonhuman (birds, monkeys, fish, plants), and parahuman (including the technologies discussed below, such as autonomous recording units and pattern-identifying software). As Michael Gallagher, Anja Kanngieser, and Jonathan Prior (2017) write

> [E]verything participates in the *sounding of worlds*, including both biotic and abiotic bodies—an exhale, the teeming of insects, the movement of fabric, a chemical reaction, the oscillation of leaves and branches,…an echo off concrete, a riot, a boat idling,…ice thawing and so forth. (p. 9, emphasis added)

However, to the extent that sound is the propagation of vibrations in a medium, those vibrations can be transduced into other formats and other sensations. Pinch and Bijsterveld's (2012) insistence that sound is always perceived under the conditions of a transduction of vibrations between bodies has become a presumption in the field of sound studies. They argue that this process of transduction "turns sound into something accessible to other senses" (p. 4), and includes other syntheses, such as data sonification and sonic visualization. Such a perspective goes well past the notion that sound is *for* any one perceiver to be perceived in any one way, thereby welcoming a more expanded definition that makes available new human / multispecies / more-than-human networks of relations and possibilities.

The concept of an infrastructure of audibility contributes to new cultural theories of sound, certainly, but does not reduce sound to the status of wave, event, object, phenomenon, or transduction. Instead, newer sound theories embrace a more multifarious, expanding, and evolving explication of the intertwining of sound, body, place, sensation, and the virtual. These understandings, in turn, join with advancements in cultural theory linked with the natural sciences. The virtual, the haptic, the affective—in short, that which vibrates beneath or above the surfaces of perception—are making their appearance within disciplinary streams beyond sound studies, where they participate in explorations of disparate sets of concerns.

To return to an earlier point on sounding, such sound theorists as Mark Grimshaw and Tom Garner (2015) are more interested in the question of sound as a *method*. For them, sound emerges from within *exosonic* and *endosonic* vibrations, enabling us to map a field, milieu, territory, or place, thereby generating a "sonic aggregate." In the sonic aggregate, when exosonic elements engage an endosonic series of associations, the emergent perception unfolds as though it were an aural Kanizsa Triangle, where form and contour may appear where none actually exist. None of this would be possible, however, without a spatiotemporal point of reference, a continuing framework of variation within the endosonic and the exosonic, and the edges that contain them. Sound, Grimshaw and Garner claim, is situated, in that "all sound exists within the acoustic ecology of the mind, of which the environment is a component" (2015, p. 74).

Another example of situated sounding is Eldritch Priest's (2018) recent contribution where he approaches sound from an "epiphylogenetic-technic" perspective, critiquing the technological objects of sonic storage and dissemination as encouraging a capitalist encroachment upon the unconscious. Extending this situated notion, Steve Goodman (2010) has introduced Alfred North Whitehead's concept of *prehension* into the lexicon of sound studies. Within a framework of understanding in which sound technologies capture and distribute dimensions of sound that were previously imperceptible, prehension subverts the established "sound-listen" dichotomy:

[Prehension] exceeds the phenomenological demarcation of the human body as the center of experience and at the same time adds a new inflection to an understanding of the feelings, sensuous and non-sensuous, concrete and abstract, of such entities. To feel a thing is to be affected by that thing. The mode of affection, or the way the "prehensor" is changed, is the very content of what it feels. Every event in the universe is in this sense an episode of feeling, even in the voice....Crucially however, the hierarchy does not imply the dominance of conscious over nonconscious vibrations. At every scale, events are felt and processed as modes of feeling before they are cognized and categorized in schemas of knowledge. (p. 23)

However, where cognitive approaches situate the mind as the generative locus for a sounding event, Goodman's (2010) project is a philosophically opposed materialist (and arguably posthuman) perspective, because it seeks to dethrone the exceptionality of the human subject by looking towards vibrational excesses. This abundance still returns us to a robust understanding of the sense, affect, atmosphere, elemental, and infrastrucutral preconditions to human and more-than-human subjectivity. For Goodman, sound is thus intertwined with the haptic, or "touching at a distance," to echo R. Murray Schafer's dictum (1993, p. 9). However, at the centre of Goodman's analysis is how such sounding affects change in the sounding object. All sounding happens at multiple, multiscalar, and simultaneous scales, and is embodied at once in-between the multiple dimensions of infrastructural audibility.

Priest (2018) extends this set of concerns to the role of the listener. He argues, albeit furtively, that the proliferation of techno-mediation has activated new *non*-listening practices of inattention. Priest clarifies this:

> Just as a computer mouse makes our body's potential to suffer repetitive strain injury a functional phase of our prehensile dealings with a technological milieu, so too do recordings make listening distractedly—that is, listening non-listeningly to the technologically occasioned abstractions of vital activity—a functional phase of our evolving auricular relations with the world. (p. 10)

Significantly here, Priest has touched on a particular transition in media theory towards the elemental, which is generating new insights. Although somewhat slow to embrace this turn, sound studies have caught on to this general enthusiasm within media theory about the elemental, atmospheric, and infrastructural sensoria that underlie our experiences, as well as the implications that those mediations have on our philosophical notions of subjectivity, objectivity, and potentials for theorizing.

Specific to sound studies, a great deal of attention has been directed towards the inaudible processes of sonic life—or as Michele Friedner and Stefan Helmreich (2012) describe, those processes that escape a more "phonocentric" perspective on sound and sound studies. We are currently on the move to produce a more inclusive field of sound studies, that encompass the breaches of the human/nonhuman divide, and to look at and feel sound and voice beyond the cochlear-centric and laryngealcentric perspectives. Such a perspective is certainly captured by John Sterne and Mitchell Akiyama's (2012) short article on sonification—a now thriving area of research—as a means of addressing the necessity to cross divides such as the visual/sonic, the human/nonhuman and nature/culture. They write that

> sonification is a particular articulation of the sonic and the nonsonic, one that points to an increasing vagueness of the borders around the audible world. Sound studies of the present moment must therefore wrestle in new ways with the boundaries of their objects. The field must let go of its axiomatic assumptions regarding the givenness of a particular domain called "sound," a process called "hearing," or a listening subject. This is not a call for a kind of everything-goes postmodernism but rather a reminder of the articulatedness of sensory technologies, sense data, and the senses themselves. In the rise of sonification, we note an increasingly forceful articulation of the senses as permeable and susceptible to transcoding. (p. 556)

Against the background of interest in the edge of sonic perception, sonification centralizes that which vibrates on the periphery of conscious awareness, while bringing into consciousness those vibrations that underlie our embodied everyday life. Sonification represents an attempt, from the perspective of media theory, to make sense of the ubiquitous media that hang on the edge of our experience and our perception. Thus, as experience opens to those peripheries, Sterne and Akiyama argue that sound studies must also be receptive to the transductive processes of interpreting peripheral vibrations.

Written as a contribution devoted to digitized sound recovery of nineteenth-century sound recordings, Sterne and Akiyama (2012) mark the shift towards temporally and spatially expanded (and likewise compressed) mediations with a special interest in the phenomena that, while beyond sensation, impose changes on our perception. Sound in the context of contemporary digital culture turns our attention towards new audibilities (or audio-abilities), but in ways that require researchers to adopt cross-sensory methods to analyze these sonic worlds. Against the cochlear-centric definition, sound need not be physiologically recognized by a normative human body (a privileged actor) to be registered specifically as sound. We are at a crossroads in sound studies, Sterne and Akiyama (and others) argue, where sound must be recognized less as an object of analysis and more for its potential as a method. Indeed, biology, geology, engineering, STS, architecture, and many other disciplines are finding value in sound as a method for understanding central disciplinary concerns of their own. These invocations make obvious that sound studies scholars can oblige this broader engagement by analyzing what sound *does* as knowledge, instead of how it is produced as an object through and of knowledge. How we come to know through sound is decidedly a question of methodological concern.

To designate the phenomenon, Sterne and Akiyama (2012) point towards the need to develop a "postsonic" imagination. As such, they embrace a sonic plasticity that puts sonic transductions into conversations with other technologies of sense and sensation. Thus, sound studies are currently breaking away from many familiar tropes, not only for philosophical reasons, but because sound and listening are becoming increasingly digitized, read, and manipulated through computation. Additionally, sound studies include new nonhuman realities that the sciences and other intellectual movements (like critical posthumanities—new materialism, affect theory, and feminist science and technology studies) are connecting with the natural and technical worlds.

Nonetheless, I proceed with caution. Sterne and Akiyama (2012) eventually reduce these new and exciting possibilities into something that recycles the dichotomies of sound and listening that they so admirably critique. They ultimately return to sound as something that is intended to be listened to: even in the face of a dismissal of the listener, they constitute the listener as the *sine qua non* of a sounding event. Here, I ask instead to abandon entirely the notions of "the sound" and "the listener"—to put listening on a transmediational spectrum. I especially question the listening experience, or the compulsion that we have to listen to better understand, and that listening is a process of colonizing space. To understand how sounding as a method eschews the preservation of sound as an experience in listening, it is necessity to assume a new radical stance with sound and sounding.

This new attitude positions sounding as enmeshed in data processing and sound-recording processes, and as purely data. Sound thus retains its shape and character as a slippery fold between the material and the virtual, enfolded in the space-time that we inhabit. But it is not meant to necessarily be listened to. Sound, instead, is part of the configuration in a network or infrastructure. It only becomes clearer in its purpose when we look at all the things around sound that sustain sounding as a creative and cultural practice. This practice resounds through our capacity to live with and through sound, even while we are not bothered to contemplate it, to adopt a hearing aesthetic instead of a listening aesthetic, or to adopt an analytic and visual perspective instead of a contemplative and sonic one.

In Chapter 4, I illustrate this perspective by drawing on the example of the microphone as an instrument of sonic intimacy that makes no such sound itself. The microphone instead opens onto an environment, or more precisely it "picks up" the environment. A simple instrument central to sound and sounding (but not so often the sole focus), the microphone is the functional infrastructure of the very possibility of sounding. Microphones open onto atmospheres: they are elemental because as they instantiate the recording process, or make the recording process possible (though they themselves do not record). However, it would be obtuse to dismiss the sonic imagination in any consideration of the microphone because the microphone picks up vibrations. What if we were to think through the microphone not as capturing data but as an opening onto data? It is precisely this gesture I want to capture in the "becoming ecological" of voice. We are accustomed in sound studies to the claim that the microphone (and more broadly the process of transduction) has removed the voice from the body. But this chapter asks if the microphone reassembles bodies into new multiplicities.

Pondering the microphone's role in the infrastructure of audibility is to consider the listener at the *postsonic* listening future. Still, it is necessary to emphasize the *placement* of microphones, which is the *pre-sonic labour that feeds a sound event.* There is something ontologically autonomous to microphone placement, like some technological affect that precedes the very possibility of a sound. The microphone offers us a glimpse into this notion of an opening that precedes a dwelling, a listening; it speaks to orienting oneself towards, in anticipation of, the sonic event. The listener does not, by definition, occupy the other end of the communication. The microphone, we will also discover, is not alone. It is accompanied by other media that feed into the cultural assemblage.

This book contributes towards a de-anthropocentrized sound studies, aiming to position an ethical embodiment that locates the openings of bodies across voicing, sounding, and placing. It adds to an enriched understanding of embodiment, sound, and science by displacing sound and the voice from the human and locating it instead in relations between human and nonhuman, biotic and abiotic entities. This rotation thus transcends the boundaries of human sounding and understanding, while also relocating these relations across timescales that are, by turns, massive or minuscule. Much of the theory for this de-anthropocentrized sound studies comes from sound-based sciences, but also from readings of contemporary media theory and cultural production, including sound art and soundscape compositions.

Chapter Overview

Chapter 2 is an individuated perspective on the voice and the body, exploring how the voice emerged in the condition of a scientific bifurcation—that is, where the experience of the voice as an event of the body shifts instead towards the voice as a collection of techniques attributable to an area of the body. It is intended as an approachable introduction to the infrastructure of audibility and the imaginary organ. Moving away from the human voice as a discourse, the chapter explores the media materiality of the early voice sciences, with particular attention devoted to the invention of the laryngoscope in the nineteenth century as a form of embodied epistemic visualization of a body's infrastructure of audibility. The vibrational folds of tissue were discovered by a handful of physiologists, but the vocal pedagogist and opera singer Manuel García was instrumental in fortifying the laryngoscope and conveying its importance to the scientific community. His characterization of good strong voices is part of the "voice-myth" that carries over into today's health and cosmetic surgery industries, and produces the image of the *laryngealcentric voice*, the imaginary organ that is now the, more or less, dominant way we think of the voice in cultural studies and media theory.

Through this particular bifurcation, the voice also became the foundation for modern acoustics; as such, an abstract science was grounded in data accrued from the diseases of the voice. In this chapter I argue that the practices of laryngoscopy produced the voice as we know it today: a concretization of vibrational practices that were attached to the value of the human body. The chapter explores this sociotechnical production as one of a *general organology*, which is taken from Bernard Stiegler's projection of the organ into biological, technological, and organizational entanglements that produce the first acoustic embodiment and concretization. It is from this particular theorization that the rest of the book attempts to think of sounding beyond human embodiment.

Chapter 3 continues with the voice but does so in the context of an *ecology* of voice relations—it also proceeds by way of a metamorphosis. Instead of concentrating on a case study, it places a variation of vocal utterances and bodies in science, art, and nature alongside one another in an attempt to capture a metamorphosis. This is done to capitulate what Gilles Deleuze and Félix Guattari (1987) describe as the *becoming-animal in voice*. The intention is to expand the infrastructure of audiblity so we can hear how it touches simultaneously on multiple scales of sounding in science, art, and philosophy. It also simultaneously produces the voice as an imaginary organ of openness and connection, a touching between multiple surfaces, utterances, and iterations, from screams to grunts to rasps to howls. It introduces the concept of a *voicescape* to capture how voices are relational and ethical, and produce more-than-human communities between animals, humans, and technologies. The chapter turns to Deleuze and Guattari's unique theory of voice as a site for becoming-animal, which offers a view of the voice as an opening onto ethical relations in place of the individual. Thus, whereas Chapter 2 argues that the healthy voice was conditioned by a sociotechnical imaginary of the voice's vibrational essence, Chapter 3 turns towards those voice pathologies as the site of a "continuous becoming" between animal, human, and technological assemblages. *Opening* ripens into a central concept in this chapter, describing how negative spaces evolve into entities, making new relations possible, grounded in examples of grunting, yawning, coughing, crying, and laughing. Concluding with a meditation on Maria de la Bellacasa's (2017) ethics of care, the chapter concludes that the voice does not represent a body, but that the voice is the body embodied.

Chapter 4 explores the infrastructures of audibility in bioacoustics field research, along with the transacoustic objects and communities that contribute to catching an event, towards the capture of the imaginary organ. Grounded in field research with bioacoustics researchers and research technicians, the chapter explores two modalities of a familiar technical sounding object, the microphone, to illustrate the ways in which data collected morph into the shape of the instrument(s) used to collect data. That is, at once capturing a world and contributing towards its creation, new research instruments combine with old to make new multitemporal worlds that fuse meaning between animals, technology, and researchers. I use the term "transacoustic community" to describe the technologies that carve out the edges of empirical data, crystallizing it in a moment of collective individuation.

In Chapter 5, I expand on the animal's voice as it encroaches upon the scientific field of bioacoustics, an interdisciplinary field bridging biological and sciences that uses sound technologies to record, preserve, and analyze large datasets of animal communications. As demonstrated in Chapter 4, it is also a world made of the meanings created through inter- and intraspecies communication. Whereas Chapter 4 examined the infrastrcuture of audibility that goes into capturing sound events, Chapter 5 asks what happens when the infrastructures of sound data are too massive to be audible to by any biotic entity. This actuality facilitates theorizing the expanding sense and sensation of a global biosphere and sonic data, and turns our attention towards the imaginary organ of the global sonosphere as well as the vibrations beyond human experience. This chapter contributes to methodological discussions regarding the longstanding questions of how researchers and scientists are implicated in the knowledge and objects they collectively produce, and how they value infrastructures of audibility on a global and longitudinal scale. This is accomplished by giving a sustained, detailed account of the science of (computational) bioacoustics—particularly how its modes of measurement allow for a new way of understanding what is involved in the decentred modes of hearing that recentre acts of listening—and by considering the nature of the relation between researcher and researched. Thus, computational bioacoustics offers new decentred ways of experiencing sound and vibration on the periphery of direct experience, which leads to the final chapter on geoacoustics and the sounding of place.

In such a context, where sound may not be listened to by human listeners, Chapter 6 most radically removes sound from sound studies in its rhythmic analysis of a place, emphasizing how the infrastructure of audibility is manifested in the vibration of place, taking up how a place produces rhythms and vibrations through multilateral timescales. Grounded in a case study of the Frank Slide (Canada's deadliest rockslide), this chapter introduces a new rhythmic and temporal perspective on disaster sites as organs on the cusp of the past and the future, and bound by a vibrational ethic of the earth, which produces multiscalar vibrations that are vastly imperceptible, enormous things of scale and pace; the rockslide is the imaginary organ that binds the sounding of place in terms of its temporal retentions and protentions. This unfolds by correlating the specific technicality of scientific research and management of disaster sites with a broader vibrational framework from within social sciences and spatial theories.

Heritage sites such as the Frank Slide are often understood as protected places that benefit the image of a sovereign nation (i.e. a "place-myth"). It is often assumed that heritage sites need protection from natural elements and from human interference. But the case of the Frank Slide is different, in that (a) it is a heritage site made out of the remnants of a terrifying disaster and (b) it is predicted to be further damaged when its ensuing rockslide follows (sometime between now and 5000 years). This makes the case of the Frank Slide an intriguing one for an interdisciplinary study, since it is made up of various overlapping temporalities, rhythms, echoes, and other earmarks belonging to the *measurement-time* of scientific monitoring, the *commodity-time* of the tourism industry, the *myth-time* of national identity, the *duration-time* of cultural memory, and the *anticipation-time* of further disaster. The analysis considers how these disparate activities contribute to the vitalization, devitalization, and revitalization of place in a way that challenges the "dark tourism" paradigm that has come to frame disaster sites. Chapter 6 thus proposes a unique synthesis between infrastructures of audibility and the imaginary organs contained within them so as to elucidate and explore how various overlapping temporalities make up the visible and invisible materials of a place. This chapter offers what Doreen Massey (2005) has called an "alternative positive understanding" of disaster sites, by describing the epistemic and cultural coordinates that make the Frank Slide an opportunity for sounding the finitude of time.

The concluding chapter synthesizes the themes explored throughout the book, especially the openness of sonic data, while pointing in the direction of data philosophy for future research in sound studies. By using sound as an example of building ethical relations between trans-species communities, this concluding chapter explores the growing divide between the perception and sensation of sound. It includes the roles of voice, vibration, sound, and oscillation as a final thought for an expanded study of scientific methodologies that use sound as it is entangled in processes of datafication.

Each chapter offers a disparate yet threaded definition of sounding according to the material, virtual, and ethical elements, and contributes to the advancement of media theory and STS by bringing elemental mediation into media theory and responding directly to current developments in the philosophy of media. Collectively, the chapters bring peripheral, elemental, and atmospheric mediation into the domain of sound studies, especially those mediations that expand our definition of life and community beyond the boundaries of human subjectivity and into new transdisciplinary theorizations of evolving techno-biological assemblages. The argument here is grounded in contemporary scientific and cultural pursuits that amplify new infrastructural embodiments of voice, sound, and sense—especially as they are manifest in bioacoustics research—to stress a general vibration within mediation, instead of perpetuating deterministic divisions between technological objects and biological organisms. I demonstrate how sonic research opens onto important areas of inquiry, even while it is opened by them, that range from shared ancestry to biodiversity preservation and bioethics. This reciprocal process produces new possibilities for imagining the concept of the body and embodiment as an acoustically material, virtually organized, vibrational mesh of forces. I argue that sounding emphasizes the importance of a general vibration that encodes human, animal, geological, and technological problematics.

The study is energized by and hopes to energize scientific methods that generate the discovery of new ways of becoming-with the collectives in which they are situated.

References

Berland, J. (2009). *North of empire: Essays on the cultural technologies of space.* Durham: Duke University Press.

Bijsterveld, K. (2019). *Sonic skills: Listening for knowledge in science, medicine and engineering (1920s–present).* London: Palgrave Macmillan.

Bruyninckx, J. (2018). *Listening in the field: Recording and the science of birdsong.* Cambridge: MIT Press.

de la Bellacasa, M. P. (2017). *Matters of care: Speculative ethics in more than human worlds.* Minneapolis: University of Minnesota Press.

Deleuze, G., & Guattari, F. (1987). *A thousand plateaus: Capitalism and schizophrenia* (B. Massumi, Trans.). Minneapolis: University of Minnesota Press.

Friedner, M., & Helmreich, S. (2012). Sound studies meets deaf studies. *The Senses and Society, 7*(1), 72–86.

Gallagher, M., Kanngieser, A., & Prior, J. (2017). Listening geographies: Landscape, affect and geotechnologies. *Progress in Human Geography, 41*(5), 618–637.

Goodman, S. (2010). *Sonic warfare: Sound, affect, and the ecology of fear.* Cambridge: MIT Press.

Grimshaw, M., & Garner, T. (2015). *Sonic virtuality: Sound as emergent perception.* Oxford, UK: Oxford University Press.

Hansen, M. (2015). *Feed-forward: On the future of twenty-first media.* Chicago: University of Chicago Press.

Helmreich, S. (2015). *Sounding the limits of life: Essays in the anthropology of biology and beyond.* Princeton: Princeton University Press.

Henriques, J. (2011). *Sonic bodies: Reggae sound systems, performance techniques, and ways of knowing.* London, UK: Bloomsbury.

Katz, M. (2010). *Capturing sound: How technology has changed music.* Berkeley: University of California Press.

Larkin, B. (2013). The politics and poetics of infrastructure. *Annual Review of Anthropology, 42,* 327–343.

Massey, D. (2005). *For space.* London: Sage.

Parikka, J. (2010). *Insect media: An archaeology of animals and technology.* Minneapolis: University of Minnesota Press.

Pinch, T., & Bijsterveld, K. (Eds.). (2012). *The Oxford handbook of sound studies.* Oxford, UK: Oxford University Press.

Priest, E. (2018). Earworms, daydreams and cognitive capitalism. *Theory, Culture & Society, 35*(1), 141–162.

Schafer, R. M. (1993). *The soundscape: Our sonic environment and the tuning of the world.* Rochester, NY: Inner Traditions—Bear & Company.

Sterne, J. (2003). *The audible past: Cultural origins of sound reproduction.* Durham, NC: Duke University Press.

Sterne, J., & Akiyama, M. (2012). The recording that never wanted to be heard and other stories of sonification. In T. Pinch & K. Bijsterveld (Eds.), *The Oxford handbook of sound studies* (pp. 544–560). Oxford: Oxford University Press.

Truax, B. (2001). *Acoustic communication* (Vol. 1). Westport: Greenwood Publishing Group.

2

Sounding Voice

Infrastructure of Audibility I: A Strong Voice

The Freiburg Institute for Musicians' Medicine, a teaching, research, and care centre dedicated to the physical well-being of musicians, works towards finding solutions for stage anxiety, assists with pedagogy and training, and seeks to learn about the potential elasticity of the best voices in the world. A key purpose is to understand the physical habits people need so as to speak in a full and healthy voice. To this end, their researchers routinely ask singers to sing as they undergo a Magnetic Resonance Imaging (MRI) scan (Echternach, 2016). The resultant anatomical and physiological moving image allows researchers to see precise measurements of the vocal cavity as it produces sound in real time. The most publicized MRI scan is of baritone singer Michael Volle singing Richard Wagner's "Song to the Evening Star" (Merkur.de, 2016). Importantly, this MRI visualization of the voice merges sound, body, and technological innovations: in essence, it is a new means of preventative medicalization that upholds the conventions of "proper articulation" and a technological witnessing to the intricate structure of muscle tissue. As noted, the

© The Author(s) 2020
M. Vallee, *Sounding Bodies Sounding Worlds*, Palgrave Studies in Sound,
https://doi.org/10.1007/978-981-32-9327-4_2

human voice arises through a complex convergence of organs and processes, produced through the medium of the body by way of a power source (breath), a sound source (vocal folds/larynx), and sound modifiers (the vocal tract); it is generally considered to be healthiest when the body is most relaxed. However, the institute found the best full body stance to be relaxed but disciplined (Gvion, 2016, p. 156), given that either muscle tension or looseness could deprive a body of the strength required for projection.

Generalizable conclusions drawn from the Freiberg Institute's research also suggest that a voice's presence is reliant on the body's opacity, that the latter should almost entirely "disappear" for the healthy and idealized voice to become the focus of attention. Indeed, relaxing the body can support vocalization in all walks of life. As Diane Miller (2006) writes in her everyday guide to professional voice and communication in the workplace,

> Relax your breathing before you get to the office by breathing through your nose. Keep your molars slightly apart with your lips closed and place your tongue tip lightly behind your upper and lower front teeth. This will allow the perfect amount of air to come in. (p. 145)

Likewise, *Persuasion and Influence for Dummies* (Kuhnke, 2012) advises that "a committed voice resonates and conveys strength of character. A weak voice indicates that the speaker's unsure about what he's [sic] saying. When you're persuading someone to your point of view, speak with authority and move with purpose" (p. 353). The infrastructure of audibility in this instance is easy to point to: it scaffolds the structure for the ideal productive voice. As an image, its corporeal extension—the body, its movement, its vibrations, and its social state presupposed by its medicalization—follows the voice. The body works for the voice, moves into the voice, follows the voice. The voice is not a byproduct of the body, but rather the body moves in tandem with the voice that produces it.

Infrastructure of Audibility II: A Weak Voice

A procedure that was once reserved for patients with vocal cord paresis has now become a cosmetic surgery available to the body interior and exterior: the voice lift is an auxiliary procedure offered to those in the "voice industries," such as performers, singers, lawyers, phone operators, and, probably, lecturers. The willing patient has a choice between two procedures: either the surgeon injects implants through the neck that bring the vocal cords closer together, or they inject fat (or collagen) to thicken the surface area of the flesh. The voice lift is increasingly attractive to those who find their voices unwittingly confessing their body's age. People who suffer from voice impairments miss more work and cost health care $11 billion per annum in the United States; notably, the voice lift procedure helps only 60–80% of treatment-seekers (Ling et al., 2015, p. 2).

Over time, vocal cords do not sag like skin or pucker like fat. Instead, they ossify. Just as cosmetic surgery visually neutralizes the senescence of age (Conrad, 2007, p. 87), the voice lift secures the body's internal fountain of youth. When the body's communicative access to the social is through a husky broken rasp, the body cannot conceal its age—in the face of this, the impact of the voice lift resides in its reconfiguring of affective materiality. The voice lift prevents auditory cues from decaying, which would otherwise avow the terrible secrets of its "resonant tomb" (Sterne, 2003, p. 287) within which it is encased, out of which it resonates, and to which it ultimately returns. Little wonder we connect the voice with the most intimate aspect of subjectivity, yet acknowledge so readily the voice as an estrangement of the body. Arnold Aronson and Diane Bless (2011) list the vocal qualities physicians must be "on the listen for" when diagnosing any necessity for vocal interventions, such as when the patient's voice seems

> asthenic, breathy, choppy, coarse, dull, feeble, flat, gloomy, grating, grave, growling, guttural, harsh, hoarse, hollow, husky, infantile, life-less, loud, metallic, monotonous, muffled, nasal, neurasthenic, passive, pectoral, pinched, rough, somber, strained, strident, subdued, thick, thin, throaty, tired, toneless, tremulous, tremorous, weak, whining, and whispered. (p. 3)

Such a list certainly gilds the lily of abnormal vocal qualities. Given the alignment of any such qualities with "weakness," the voice lift has become a significant part of the beauty industry—a means of making the body's manifestations resonate with its visual ones. Since 2004, a select few with the necessary funds have opted for this vocal surgery (Saner, 2012). And some have instead pursued voice training to lubricate the vocal folds; in either case, they aim to achieve an auditory mask that does not betray the body's age (apart from what might be visually apparent). These intervention strategies index the auditory as among the possible ailments in the body. The voice demarcates a space in which the visual and the audible intertwine with one another, overlaid with subjectivity and with power. Clearly, those people who live with the calcifying voice that comes with age and/or disease cannot mask it so easily as they might a visible symptom, as this intervention highlights. In a culture such as ours, where age can be unconsciously connected to weakness, any possibility of obscuring a so-called weakness is highly, if not spuriously, valued.

Here it is relatively easy to see an opposite effect of the infrastructure of audibility that produces the human voice—how the voice is produced, as an imaginary organ, as a public disclosure of the body's place in the social world. Without the infrastructure of audibility, the voice stands as a secondary character to the body, instead of one of the many effects, sociologically speaking, of having a body which produces imaginary organs that extend into the social. These circumstances tell us that infrastructures produce multiscalar bodies, a range of variations on a theme; if the voice is affectively material, it is so in a continuous variation of bodies that are considered well and bodies that are considered unwell.

In this chapter I argue that the voice is real because it is *an effect* as well as an *affect* of the body; as such, I posit the first imaginary organ of the book: the voice as an imaginary organ, made real under physiological, technical, and social convergences that construct the voice as persistent, yet ephemeral. However, while the voice is a construct of the imagination, it is *not* illusionary—but its realness does not derive primarily from its being representational of a natural occurrence. I mean merely that the voice is real through how it produces sociocultural effects such as those described above. In basic terms, the voice resonates: "vox" and the Latin "sonus" are its etymological antecedents, referencing the tone and expression that arises between two objects in contact. But a voice always resonates in and with some *place*. Thus, a voice is not so much about the subject of the sounds it makes; it also speaks to the place the voice and the body co-inhabit. Following this thread, this chapter and this book more broadly will explore the role of the voice sciences and acoustics in producing the voice and the place of sounding.

Theorizing a Voice Between: Or, How Is the Organ Imagined?

There has been a good deal of excitement around the emergence of a "voice studies" field, which calls unanimously for an expanded study of the voice from a range of interdisciplinary perspectives, but ones which confounds exactly what it is we listen for or to when in the presence of another's voice. For instance, in his call for an "expanded capacity for oral expression," Brandon LaBelle (2014) proposes a theory for the voice that stresses the analytic import of the mouth as a governing medium for concrete relations. Writing that the voice/mouth apparatus represents a body's capacity for social resonance, he claims that

> The mouth is thus wrapped up in the voice, and the voice in the mouth, so much so that to theorize the performativity of the spoken is to confront the tongue, the teeth, the lips, and the throat; it is to feel the mouth as a fleshy wet lining around each syllable, as well as a texturing orifice that marks the voice with specificity, not only in terms of accent or dialect, but also by the depth of expression so central to the body. (p. 1)

I do not suppose that LaBelle (2014) wishes all of us to listen into one another's mouths. Rather, more generally, we cannot consider a voice as anything other than the channels through, and with which, it resonates. We can only attain a conceptual view to the voice by taking into account the infrastructure of fleshy chambers that contain it and express it. LaBelle's orientation towards the mouth is an especially inventive method for framing the voice as a boundary-making object of the self that serves to maintain it, while also contributing to his broader research programme on how sonic phenomena reconfigure lines and borders between embodiment and social relations. Through this lens of what we could call an *infrastructure of orality*, the voice not only allows subjects to locate and identify one another, but it also *places* them in a unique sense. By considering the site of the voice in the mouth, LaBelle reminds us that such places are malleable entities, that the voice vibrates between things that deserve analysis: organs, bodies, selves, socialities, cultures, and so forth. That is to say, as the voice crosses the threshold of the mouth—the container of all affective and linguistic meaning—it demands its seat in a place of resonance: it places, is placed, in place.

More generally, LaBelle's (2014) work suggests that sound propagates as an effect of disequilibrium, a "restlessness" of a sort, since sound indicates the movement of objects in contact or friction with one another, no matter their size: hand on table, mouth on reed, foot on ground, and the like. Thus, to return to the voice, the mouth always reminds us that the body is in motion in the sense that it contains all those social and physiological codes that bridge the body to the outside. As LaBelle notes, the voice is "an essential means by which the body is always already put into relation" (p. 1). This demands of the listener the responsibility to anticipate and listen for the presence of voice, however unstable a force it may be. Furthermore, the voice points to the restlessness of bodies and the resistance of bodies to conform to the notion of a static object. A voice is never, according to this definition of sound, a *nature morte*. To echo Deleuze and Guattari in *A Thousand Plateaus* (1987), a voice is always many voices, or "a *machinic assemblage* of bodies, of actions and passions, an intermingling of bodies reacting to one another" (p. 88). Their perspective strongly resonates with LaBelle's position, which I will return to in Chapter 3. We can take this to mean that the voice conjoins bodies—it is not an object which passes between bodies. We might extend LaBelle's thought by pondering that the voice *happens* as a connective tissue between bodies, an affective relation that makes meaning possible. To critically engage the voice, and to place it appropriately, we must think through the infrastructure of audibility as an *ecological* infrastructure.

Claims that the voice marks a permeable boundary between self and other have faced other challenges. Milla Tiainen (2013) writes that the "new materialist" paradigm disavows such a split between body and voice by favouring a more inclusive and immanent "emergence" perspective on the voice. Similarly, the editors of *Voice: Vocal Aesthetics in Digital Arts and Media* (Neumark, Gibson, & van Leeuwen, 2010) claim that in the context of digital media, the "fundamental paradoxes of voice—embodied and moving between bodies, sonorous and signifying—have become even more complex as voice, always/already culturally (and politically) mediated, is remediated and remixed" (Neumark et al., 2010, p. xxix). Likewise, Nina Eidsheim (2015) asks for a "reawakening" of the senses through a consideration of voice, not as sound but as the ongoing exploration of our own understandings of sound in vocalization. As Matt Rahaim (2019) writes in *The Oxford Handbook of Voice Studies*:

> Unlike an arbitrary act of willful fantasy, working within a vocal reality includes the possibility of being surprised: a voice can show up quite differently than we think it should. But no "voice" shows up for us as real in the first place without the infrastructure, situation, habits, and practices that make it available. (p. 26)

It appears as though we are in another interdisciplinary turn with the recently published edited collections on voice studies that explicate the voice's liminalities, relationalities, and embodiments (Eidsheim & Mazzei, 2019; Thomaidis & Macpherson, 2015). This turn includes reconceptualizing voice within the context of Deaf culture (Levitt, 2013), temporal in-betweenness (Järviö, 2015), puppetry (Mrázek, 2015), and displacement (Chatziprokopiou, 2015; Di Matteo, 2015a, 2015b). Further explorations continue with such voice- and body-related topics as "resonance" (Sholl, 2015), "vibration" (Dyson, 2009), and "echo" (Vallee, 2017). In effect, "voice studies" represents a non-unified field and a profusion of perspectives, including those researchers who do the following:

- Describe the "affective materialities" of voice by proposing the possibility for incorporating its timbre, tone, duration, and pitch into discourse analysis (e.g. Kanngieser, 2012).
- Theorize the voice as the figuration of embodied uniqueness (e.g. Cavarero, 2005).

- Consider the voice a disturbance and a blind spot that is at once a part of, yet apart from, subjectivity (e.g. Dolar, 2006).
- Approach the voice as the filtration of complex epigenetic processes (Blackman, 2016).
- Demonstrate the crucial link between voice, identity, and subjectivity (e.g. Blackman, 2000, 2010, 2012; Mazzei, 2013; Mazzei & Jackson, 2012; Mazzei & McCoy, 2010).

In concert with these developments, I am interested here in tracing the voice's imaging as an object of technical and social innovation, and thus have a corresponding interest in the bits of matter that have been scientifically identified as making up the voice: the configuration of diaphragm, lungs, trachea, vocal folds, pharyngeal cavity, tongue hump, velum, nasal cavity, oral cavity, nasal sound output, and oral sound output. My claim is that the voice is an imaginary organ in the sense that it has come, through the technologies and practices that make up its infrastructure of audibility, to be an effect built from the imagining of its causes—an object that is real but not actual, while nonetheless an object that structures our conception of human volition. The present approach to voice imaging is sensitive to the epistemological and practical effects of medical and scientific discourses on the body, and their related discourses, which trace the contours between voice embodiment and voice estrangement. Voice estrangement, in conventional wisdom, is taken as something of a fissure that needs mending, where voice is framed as the most fundamental testimony to one's corporeality, and one's presence and co-presence with others. This is a common assessment that takes the voice as a sort of entrance to the self; in the flesh, this is endowed with the capacity for enunciation and utterance (Kanngieser, 2012).

My conception of voice asserted here is intended to explicate the connection between action and practice. Central to this position are issues concerning the visualization of the voice through the larynx and those techniques of visualization that bring the body into focus as an object of knowledge. This orientation unavoidably engages with the affective materialities of the voice. Thus, I dig deeper into the body, beneath LaBelle's (2014) mouth-centric discussion of the voice, to consider the recesses of flesh and the internal workings of the larynx and the voice that had eluded philosophers and physicians for centuries. In the remainder of this chapter, I will take up affect and the affective materialities of voice as they are framed historically through imaging technologies and scientific discourses, beginning with nineteenth-century alignments of the voice with the body's aliveness and the understanding of voice as a vibrating intermediary between internal and external relations.

This chapter considers voice imaging in the context of a bioethical imperative to expand the boundaries of human communication. My claim that the voice is an imaginary organ is grounded in and grows from Bernard Stiegler's "general organology," a conceptual apparatus for understanding biological, technological, and organizational entanglements, and which I will explicate following the empirical data. Stiegler's (2004) description of organs refers to the organs in the human body, but does not necessary encompass the voice, since the voice represents an energy that arises between such organs. Organs, then, for Stiegler, are different from organizations, which refer to the social milieux that are bound by technical considerations, such as the medical industry's participation in technological innovation. Finally, within the context of organizations, Stiegler inscribes organs with meaning according to the techniques used to isolate and individuate them. Elsewhere, Stiegler (1998) sees a more socially conventionalized understanding of technology as "technics," which embraces the social, cultural, and industrial innovations behind technical progress, and the expanded understandings of embodied objects. In light of Stiegler's general organology, I will demonstrate that, as a cultural image, voice is a process through which a subject "edges" itself into representational frameworks—a process which transgresses the boundaries of technological, biological, physical, psychological, social, and cultural frameworks. To that effect, the voice is an imaginary organ.

The human voice has long been linked with the body, health, and the capacity for effective and articulate communication. Hippocrates first observed that the human voice resides deep in the body and the trachea, arising from the air that a body inhales into itself, though he considered the tongue and the lips to do the work of articulation (Baron & Dedo, 1980). Without the tongue, he conceded that there would be no articulation, and without efficient articulation there would be no effective communication. It was crucial for a functioning society that the body be physically able to emit clear sounds. It is with Aristotle, then, that we first see the connection between voice as a capacity to express a political community's values and humans as the "animal of the polis" (quoted in Cavarero, 2005, p. 184).

But the voice is, as LaBelle (2014) writes, localizable and ultimately identifiable, speaking to the malleable place that both the speaker and the voice occupy. It is also a historical and technological coordinate. I call this the "laryngealcentric" voice, the voice brought through technics into the light of science. This voice emerges via scientific experiment and practices. Of interest here is a particular bifurcation: The voice sciences separated the voice from the body of its containment. It was lost in the excitement of its discovery, and that excitement was, in turn, converted into the foundations of a science of vibration, spearheaded by Helmholtz (1885). In other words, the voice became a knowable object by virtue of its discovery through the technical tools that isolated the voice in the throat. This approach is indeed useful for the voice sciences and those patients and individuals who live with voice disorders, but it neglects the associated epistemic consequences. All those sounds of the voice, those secondary qualities of the voice, are downplayed as signals of something "wrong" in the tissues—problematic epistemic entities to be isolated. Curiously, the laryngoscopy, a route towards the medicalization and individualization of the human voice that is decidedly part of the voice sciences, came to fruition under the guidance of a singing teacher and a music theorist.

The Laryngealcentric Voice

Upgrading the Infrastructure, Discovering the Voice

One aspect of vocal bifurcation is the invention of technologies intended to carry light into the openings of a patient's body. The first such device was Philipp Bozzini's (1807) Lichtleiter (loosely translated as "light conductor"). The Lichtleiter, though never manufactured and put into use, was designed to illuminate a variety of orifices using a 33-cm long and equally wide instrument equipped with mirrors, candlelight, rods, balls, tethers, and tools for wedging apart fleshy closures in orifices and passageways, all with the intention of bringing the body closer to the eye of the examining physician. While Bozzini did not mention the voice in his 1807 patent that described the Lichtleiter, his idea of illuminating cavernous pathways of the body for the return of an image to the observer resonated through decades with the invention of the laryngoscope. He wrote of the benefits of the light conductor in rigorous detail—describing the turns the tubes take through the cavity walls of the nose, mouth, throat, anus, and vagina. His celebration of the machine and its associated techniques nods to the entire faculty of understanding, in which a representational image emerges from within the entangled folds of technics and tissue: "When several senses are directed towards an object," he wrote, "the more clearly it becomes our imagination. All representations which can only be attained by the sense of the face have been, for the most part, uncertain or lost by the caves and spaces of the living animal body" (Bozzini, 1807, pp. 15–16, author's translation).

In his short section, "Light Line for Oblique Angles," Bozzini's (1807) concern with illuminating dark, enfolded passageways for scientific measurement complements other emerging ideas of his day about technology and vision, such as relief from the spectre of illusion made possible by optics and technics, especially concave mirrors. "For the scientific imagination," he wrote, "telescopes, microscopes, and other optical technics laid open the infinite in both directions" (Bozzini, 1807, p. 17, author's translation); however, the internal contours, passageways, and sheaths of the incandescent body still awaited a new understanding of the body as a labyrinth of folded tissue and touching spaces. A significant potential of the Lichtleiter was its capacity to visualize the movements of the living body. Bozzini wrote that to judge the human body, one must not leave the imagination directly and indirectly to each vessel; otherwise one imagines the idea, and not the living, animal body. If we imagine these movements, we know that they are diminished in some systems, increased in others. "Thus," he concluded, "through the application of the face to the inner cavities and interstices of the living animal body, the medicine becomes more perfect in all its branches, and these, with a higher degree of certainty, return to themselves under their own self" (p. 7, author's translation).

Bozzini's (1807) description frames the human body along the lines of logical positivism, in that the light conductor was capable of producing a new norm for discovery, one that connected orifices to the entirety of the human body and to the corpus of knowledge about the human body. While tracing this link to Bozzini is not an exhaustive history, it is certainly one route by which an ideal voice was constructed; these notions continue to resonate today, incorporated into objective science about the voice.

Bozzini's design was guided by a penchant for optical mechanization, towards a technical imagination of the body that denied the investigator's perceptual accuracy without the assistance of technical imaging unless it was accompanied by a desire for judgement about the body. Indeed, his enthusiasm upholds a normative principle in scientific discovery. For René Descartes, for example, the eye was the container of perception, unless vision was clearly being directed by optical mechanisms and the scientist's patience. This view does not deny the existence of a material world, but rather suggests that the material world is one suspended by unfastened images of a reality grasped primarily through unreliable sensory experience. Even with a material assumption, which Descartes assumes we *all* assume, sensory experience cannot reveal the constitutive properties of the material world. In the famous example of the melting wax from *Meditation One*, Descartes insisted that one cannot perceive the general category of wax as it undergoes transformation. One instead must rely on systematic observation and a judgement that the condition of wax is constant all the while it melts. Sensation, Descartes argued, was an unreliable mechanism for gathering knowledge about the world—perception was more reliable, superseding sensation as a categorical predicate. All aspects of life and the universe were subject to unifying laws of nature, from the depths of deep space to microorganisms to the inner recesses of the animal body.

Illuminating the Voice

Benjamin Babington's Glottiscope proposed a solution to the issue of illumination by using the sun's reflection on a handheld mirror to shine past a patient's supraglottal and pharyngeal tissue (Pieters, Eindhoven, Acott, & van Zundert, 2015, p. 5). This solved the issue of a light source, which had prevented earlier explorations of the voice, from Hippocrates to Avicenna to Paul of Aegina, whose musings on tracheal intubation could never be enacted on a living patient with much success (see Younis &

Lazar, 2002). But because Babington never published on the success of his Glottiscope, it was up to his successors to write on their memories of the instrument: "It must be remembered that before the days of the laryngoscope the larynx was for all intents and purposes an internal organ" (Wells, 1946, p. 443). For physicians, those cavities of the human body interfered with medical diagnosis; for proper diagnosis, the physician was required to see the throat produce sound in real time, not in the throat of the dead. To capture the voice not only in light but in movement would have been a milestone in discovering the interior of the human body.

Bozzini's and Babington's designs and innovations are important because they brought light to the body and brought the body to life, even though the devices themselves did not gain much traction. The body was understood—within the context of medicine, biology, and scientific discovery—as a living organism caught in the struggle against the invisible hand of death, while subject to an optic fascination shaped by technical means of observation. Illuminating the dark folds that contain the voice is a way to understand how the voice entered a discursive framework of vibration, since nobody was certain what made a voice vibrate or resonate with the body cavity. Illuminating the body and discovering a process by which to light up the body while sensing it move, led the voice to be interpellated as a type of already-embodied prosthetic. Indeed, as James Rush wrote in *The Philosophy of the Human Voice* in 1827, "The art of speaking well, has, in most civilized countries, been a cherished mark of distinction between the elevated and the humble conditions of life, and has been immediately connected with some of the greater labours of ambition and taste." The challenge of science to capture the voice is certainly apparent. While Bozzini and Babington had succeeded in illuminating the caverns of the body, Manuel García is credited with bringing the image of the voice together with the vibrating folds of the body, with his invention of the laryngoscope. Prior to these innovations, the scientific context for studying the voice was rather piecemeal as James Rush's (1827) description makes clear:

> They [scientists] have removed the organs from men and other animals, and have produced something like their natural voices by blowing through them. They have inspected and named the curious structure of the cartilages

and muscles of the larynx, with the absurd purpose to discover thereby the cause of intonation, when they were ignorant of the very forms of that intonation. In short, they have tried to see sound, and to touch it with the dissecting knife — and all this without reaching any positive conclusion, or describing any more of the audible effects of the anatomical structure, than was known to rhetoricians, two thousand years ago. (p. iv)

With the technical capacities in place, the voice's discovery hinged on one further and particularly important factor: real-time movement. The invention of the laryngoscope by voice teacher Manuel García in 1853 (Bailey, 1996) thus contributed to a new scientific culture around the study of the human voice. Along with other inventions like the stroboscope, the laryngoscope allowed for non-surgical tracheal intubation and inspection, and supported the nascent medical field of laryngoscopy. García (1881) wrote the following about his discovery:

One September day, in 1854, I was strolling in the Palais Royal, preoccupied with the ever-recurring wish so often repressed as unrealizable, when suddenly I saw the two mirrors of the laryngoscope in their respective positions, as if actually present before my eyes. I went straight to Charrière, the surgical-instrument maker, and asking if he happened to possess a small mirror with a long handle, was informed that he had a little dentist's mirror, which had been one of the failures of the London Exhibition of 1851. I bought it for 6 francs. Having obtained also a hand mirror, I returned home at once, very impatient to begin my experiments. I placed against the uvula the little mirror (which I had heated in warm water and carefully dried); then, flashing upon its surface with the hand mirror a ray of sunlight, I saw at once, to my great joy, the glottis wide open before me, and so fully exposed, that I could perceive a portion of the trachea. When my excitement had somewhat subsided, I began to examine what was passing before my eyes. The manner in which the glottis silently opened and shut, and moved in the act of phonation, filled me with wonder. (pp. 197–198)

In "Observations on the Human Voice" (presented to the Proceedings of the Royal Society of London in 1855), García demonstrated a profoundly simple laryngoscope made of a small mirror attached to a long handle, which illuminated the throat using the light reflected from the sun. The illumination revealed a complex network of complementary movements that contracted and expanded as if a collection of tissues, and his descriptions were just as layered. Regarding the movements involved in taking a breath, García (1855) wrote:

> [T]he arytenoid cartilages become separated by a very free lateral movement; the superior ligaments are placed against the ventricles; the inferior ligaments are also drawn back, though in a less [sic.] degree, into the same cavities; and the glottis, large and wide open, is exhibited so as to show in part the rings of the trachea. (p. 218)

García wrote plainly that the voice is attributable to a vast network of ligaments, tissue, cartilage, and elastic sheaths whose sole purpose was producing vibration from within the cavity of the body. He described how the voice *explodes* in air through ligaments and tissue: "The voice is formed in one unique manner—*by the compressions and expansions of the air, or the successive and regular explosions which it produces in passing through the glotti*" (p. 221, original emphasis).

In physician John Windsor's reflective commentary on Manuel García's discovery, he commented that García's 1855 demonstration marked the first detailed analysis of the voice's inner mechanism, "having first made an extended series of examinations of the healthy larynx" (Windsor, 1863, p. 209). García's demonstration, grounded in singing techniques, was an elaborate portrayal of the involuntary movements involved in the production of voice. Early in García's text, he suggested that three parts of the voice, each with unique timbres and textures, correspond with different pitch ranges: the chest for low, the falsetto for middle, and the head for high. García argued that these registers produced different images of the voice in the illuminated mirror of his simple dentist's mirror, which was used for displaying the vibrations of the larynx and the vocal folds. But the success of the laryngoscope depended almost entirely upon the self-examination of the researcher.

Where previous studies failed, García's succeeded because he was his own patient. In 1867, physician Antoine Ruppaner (1867) remarked in the *New York Medical Journal* on the profundity of this moment:

Standing with his back to the sun, he held a looking glass in his left hand before his face, the sun's rays were thus reflected by the glass into his open mouth. Then he introduced a dentist's mirror—previously warmed—into the back of his mouth and by placing it at a proper angle, he was able to see the reflection of his larynx in the looking-glass. Auto-laryngoscopy was discovered; it was an indisputable fact. His feelings—when he saw for the first time how the vocal cords acted whilst he was emitting sounds—were described to his son Gustave García as follows: "I was most fortunate, for I succeeded almost at once in seeing with my own eyes what I had conceived so long; it gave me such a turn that I felt on the point of collapsing." (p. 5)

Living the Voice

While the illumination of the body has been established as a commonplace practice that frames an organ as possessing explanatory power (the cause + effect), the voice was different given that it was a force that arose from the infrastructure of a flow and an obstacle: breath (air, a function) against the vocal cords (tissue, membrane, vibratory and enfolded flesh). The human voice, unlike other scientific discoveries of organs and their functions, was the effect of smaller organs, such as the larynx and the mouth, as well as larger organs, such as the chest, the gait, and the posture of the entire body. The human voice was identified according to the delicate aggregation of smaller and larger moving parts, and thus required an observant eye. As Tim Scott (2010) observes, the laryngoscope is an example of an instrument that goes into the body and is followed closely by a skillful eye and hand, and was created by "inventors [who] so completely embody their invention in their thoughts and actions that others cannot emulate their dexterity in its application" (p. 56).

Following the discovery of the voice inside the body, the binary approach to humans' vocal capacity was next marked by conceptualizing the voice into physical and physiological acoustic signal systems. While the *physical* was aligned with the assemblage of moving tissue in the human body, the *physiological* referred to the mental associations of those physical impressions on the listener. García, a trained singer, grounded his discussion of "voice quality" in the sciences, which influenced other voice coaches and writers to follow a path of vocal pedagogy and professional voice training that was based on the physical and physiological discoveries of the voice in operation. This is not to suggest that the technology of the laryngoscope necessarily determined the discipline of voice training. Rather, it opened a set of possibilities to (a) discover multiple folds working simultaneously and (b) facilitate, with the simplest instrument, a conversation belonging not only to scientists, but to musicians and to pedagogues as well. Thus, science and singing were brought together by a common interest in the tissues of the human body and the embodiment of those tissues. In terms of scientific value, a voice specialist was now of benefit to a wider array of professional singers, lawyers, teachers, and others who spoke in public. Soon thereafter, a wide variety of professionals stepped forwards with voice complaints, and a thriving industry developed out of García's discovery.

García's discovery of the voice was based on a curiously exteriorized, live, image-based form, founded on a series of involuntary movements transmitted through a delicate combination of mirrors, sunlight, posture, and bodily contortion. And while García's original 1855 lecture was slow to become cited, some physicians later experimented with García's method, giving him credit and eventually establishing this as a successful examination. Johann Nepomuk Czermak (1861), for instance, improved on the laryngoscope by constructing different sizes of mirrors and demonstrates its efficiency in hospitals across Europe. While Czermak encouraged his students to attempt laryngoscopies and rhinoscopies on deceased bodies, his main interests were in laryngoscopy's access to the *live body*; repeatedly, he wrote on the superiority of observing the inner workings of the living. The implications of such a concern were primarily in the areas of physiology and practical medicine: the human voice allowed physicians to observe how the body manifests itself *beyond itself* through both involuntary and voluntary movements. Such a binary contradiction reflected a general fascination with the phenomenon of *autoscopy*, which is to see oneself in a peculiar and partially recognizable form—part of, yet apart from, one's own body. The voice was the apogee of such an idea.

The laryngoscope was central to the discovery of a healthy body and how, if undamaged, it generated a series of sounds that were deemed appropriate, by moving air through obstructions—that is, the larynx. But the physician also had access to a wide range of embodied sounds that spoke to the damage of the patient. The science of the lived body also required interaction with the patient, and Czermak (1861) was particularly interested in diagnosing physiological conditions by asking the patient, before the mirror, to make sounds. Upon hearing these sounds, he would lubricate their vocal folds so as to compare healthy versus unhealthy vibrations. Czermak also used sounds of patients, such as "eh!" and "ah!" to access different configurations of openings in and through the patient, and to receive a clear view of the larynx; sounds were thus meant to "unconceal" obstructions in the flow of the body, and differences between vowels sounds sparked new interest in the voice.

Many would follow García and Czermak in the perfection of the laryngoscope. In the context of laryngoscopy, the larynx was considered an access point to the diseases that were heretofore unrealizable without live imaging technologies. Medical lecturer Prosser James (1885) wrote on the importance of liveness and the image of the voice, noting that

> the image at which we gaze in the mirror differs indeed so much from the organ as dissected after death that although familiarity with its anatomy is necessary for various purposes, the appearance presented during life is of far greater importance. It is, then, with the laryngeal image we are just now concerned. (p. 56)

This is all acceptable for laryngoscopy, but Baron Hermann Ludwig Ferdinand von Helmholtz (1885), whose ophthalmoscope Czermak had adopted for the perfection of the laryngoscope, was instrumental in connecting the voice and its live vibration to physics in its earliest years. Helmholtz agreed that the discoveries of the laryngoscope could account for the involuntary movements and vibrations of the organs comprising the larynx, but could not account for the minute changes in vowels that were produced mainly through the organs near the mouth, and resonated back into the pharynx as well as the larynx. As a response, Helmholtz produced, with Friedrich Fessel, a synthesizer for synthetic vowel production (see Pantalony, 2009, p. 22). The purpose of the synthesizer was to demonstrate that vowel sounds stood as an example of timbre, the quality of a sound. Helmholtz conceded that the difference in quality between sounds was because of the overtones produced by changing the shape of the mouth. As initial proof, Helmholtz produced different vowel sounds overtop an open piano with the string's dampers off, noticing that different strings vibrated in sympathy with different vowel sounds. This led him to commission the design of the vowel synthesizer: cylindrical resonator tubes were placed with covers at a distance from a series of tuning forks, each of which represented a tone in the harmonic overtone series, and each of which vibrated in varying intensity in correspondence with how the cylindrical resonator tube's entryway was covered (like the mouth forming different vowels allows varying degrees of air pressure to escape). Changes in the shape of the mouth produced varying partial tones that resulted in a difference in quality between sounds. But it is less the science of overtones or partial tones that is relevant to this discussion, than the question of coming full circle to the imaging of live vibration to demonstrate the voice.

Much of Helmholtz's (1885) discoveries relied on an interactive environment, governed by intimacy with and between objects. It is well-known, for instance, that singer Emma Seiler conducted the singing experiments with Helmholtz to make the piano's undampened strings resonate with her professionally trained voice, as well as participating in preliminary studies testing partial tones by holding tuning forks against her mouth as she changed vowels throughout a sustained sung pitch. Helmholtz also reportedly performed laryngoscopy examinations upon Seiler, remarking that her vocal cords resonated with great precision between tones. Meanwhile, such a precision could only be captured in an environment that demanded a certain intimacy between observer and observed, as the above examples in laryngoscopy have already demonstrated. In her own writing, Seiler (1884) also observed that Helmholtz had carried a more physiological view of intimacy in sound:

> Music, above all the other arts, finds the earliest and most universal recognition, and almost every one listens to it with pleasure. Helmholtz says that music is much more intimately related to our sensations than all the other arts put together. Tones touch the ear and are instantly felt to be agreeable or disagreeable, while the impressions of painting, poetry ... upon our

senses must be brought to our consciousness, and be judged of there by comparison. (p. 147)

Helmholtz's interest was thus in the intimacy between sounds, their sympathetic vibrations and their resonances. One of Helmholtz's imaging technologies was the vibration microscope, which was intended to capture the shape of a string in motion, more precisely a *lissajous* pattern that manifested in a triangular shape. We see here that vibration becomes motion becomes image, and while they are separate categories, in the scientific imagination they are all intertwined with one another and difficult to parse out. Thus, another point of connection worth considering is that of the light in that visual manifestation. Indeed, a new assemblage of the voice, the principle of it, was manifest in new technological apparatuses that not only allowed researchers to see it "in action" but to mimic its movements in the very substance that brought it out of the dark in the first place: in resonance.

Helmholtz is known for his theory of resonance, which uses the voice as an empirical field, and is informed more broadly by his theories of acoustics. The human voice, Helmholtz (1885) observed, could produce a fundamental pitch by singing and sustaining a tone, but simultaneously produced overtones or partial tones that resonated with that fundamental sound, and were only barely perceptible to the human ear. Such a production required singers to articulate a properly sustained tone through good posture so that the voice could resonate clearly in its vocal tract. To perhaps simplify, Helmholtz conceived of the voice as an event between the voice source resonating with the vocal tract, and extending into the space that lies beyond the body of articulation and resonance. The subject's capacity for articulation thus became a focal point, since it was relatively clear that the human voice was not necessarily an acoustic phenomenon as much as it was an articulatory one—as the subtleties in sounds were produced in the slight variations between sounds produced through the parts of the mouth. Helmholtz explained:

In order to understand the composition of vowel tones, we must in the first place bear in mind that the source of their sound lies in the vocal chords, and that when the voice is heard, these chords act as membranous tongues, and like all tongues produce a series of decidedly discontinuous and sharply separated pulses of air, which, on being represented as a sum of simple vibrations, must consist of a very large number of them, and hence be received by the ear as a very long series of partials belonging to a compound musical tone. [...] If they do not close perfectly, the stream of air will not be completely interrupted, and the tone cannot be powerful. If they overlap, the tone must be cutting [...] as those arising from striking tongues. On examining the vocal chords in action by means of a laryngoscope, it is marvelous to observe the accuracy with which they close even when making vibrations occupying nearly the entire breadth of the chords themselves. (pp. 103–104)

Helmholtz (1885) had considered the ear as the prime receptor to the sensation of sound, although it becomes obvious throughout his *On the Sensation of Tone* that sound opens up new ways of relationality and new, relatively affective ways of conceiving bodies. For him, the suspension of a tone was exemplary of how the voice touches at a distance: the voice resonates in air within the body of the articulator, while that air travels through exterior space and into the ear of another individual. But this is not a lasting contact. Instead, he argued that oscillations produced contact between bodies at irregular periodic motion, which he described as "one which constantly returns to the same condition after exactly equal intervals of time" (p. 13). Such periodic examples include the "motion of a clock pendulum, of a stone attached to a string and whirled round in a circle with uniform velocity, of a hammer made to rise and fall uniformly by its connection with a water wheel" (p. 8). All periodic motion cycles at a certain number per second (from 30 to thousands), but Helmholtz discovered far shorter vibrations for the tones articulated through the voice, thus making for a significantly more lasting rapid oscillating motion. It is the change that is introduced to another body which is the point for Helmholtz's theory. Finally, Helmholtz's idea for vibration was that the vibratory essence of relations between bodies was a visible phenomenon, even if by only technical means. He wrote:

The musical vibrations of solid bodies are often visible. Although they may be too rapid for the eye to follow them singly, we easily recognize that a sounding string, or tuning-fork, or the tongue of a reed-pipe, is rapidly vibrating between two fixed limits, and the regular, apparently immovable image that we see, notwithstanding the real motion of the body, leads us to conclude that the backward and forward motions are quite regular. (p. 8)

Less was the voice a sound than it was a form of touching, a contiguity. But it was considered also interrupted. Image and resonance, not image and movement. Here vibration is relational, but it is also only relational between very specific bodies such as tuning forks and resonators tuned at the same pitch, or the human voice and the human ear, which are capable of producing and hearing the same range in tones.

This brief historical sketch indicates that we cannot attribute a voice so readily to the body of the individual, since knowledge of its production grew outward towards different sciences and different cultural domains. Thus, the voice deserves a cultural theory that considers its ethical and de-individualized imagination over and above its representational role in serving the body. The voice, instead, is deserving of a more intimate engagement between movement, image, and affect. Gilles Deleuze (1986) observes that the imaging of vibration represents a broader intimacy between movement and image that became important with the turn of the twentieth century. Indeed, this intimacy within science and technology had made it increasingly difficult to imagine one without the other in all other realms. Deleuze argues that this coupling of vibration and matter, of movement and image—of images that move and movements that are imaged— generally reflect a world that is increasingly pervaded by such a confluence of forces. This did not create a confusion in images so much as it fortified a certainty: that images moved as much as movement was imaged.

The Voice as Embodied Index

It is an inveterate belief in cultural theory that the voice simultaneously indexes the body as it escapes the body, and as a consequence is a vibrating intermediary between the body's first-order physical presence and second-order representational presence (Barthes, 1977). The voice is thus an achievement infused with dual value for people seeking to capture their own "inner voice," a lighthouse of sorts, illuminating a hitherto undiscovered confidence with which to inhabit the world. Without the voice, such as in the case of stage fright, loss of confidence, or bodily traumas, our "whole field of possibilities" collapses, as Maurice Merleau-Ponty (1962, p. 188) described it. We might wonder what the invention of the laryngoscope can offer in such instances, when an individual is so estranged from their own voice that it bespeaks a kind of problematic embodiment. Conceptualizing the voice as an embodied force implies a listening and speaking, and summons multiple presences and multiple requisites to listen (see Back, 2007). This is a theme taken up by Steven Connor (2000) on the intertwining dimensions of listening and voice:

> If, when I speak, I seem, to you, and to myself as well, to be more intimately and uninterruptedly there than at other times, if the voice provides me with acoustic persistence, this is not because I am extruding or depositing myself with my voice in the air, like the vapour trail of an aircraft. It is my voice of my self, as the renewed and persisting action of producing myself as a vocal agent, as a producer of signs and sounds, that asserts this continuity and substance. What a voice, any voice, always says, no matter what the particular local import may be of the words it emits, is this: this, here, this voice, is not merely a voice, a particular aggregation of tones and timbres; it is voice, or voicing itself. Listen, says a voice: some being is giving voice. (pp. 3–4)

This perspective of the voice shares much in common with contemporary theories of infrastructure and mediation. John Durham Peters, for instance, opens *The Marvelous Clouds* (2015) with the insight that communication secures a connection between bits of matter: be they an entanglement of family members, telephones (all the wire and fibre-optic cables), countries and continents, or satellite systems. All of these persist through ongoing communication. In much the same way, so does the voice. Connor's (2000) voice is a voice that produces itself in a finite time and a finite space: it is localizable, to return to LaBelle's (2014) insights, situating the individual to whom the voice belongs. The voice is thus the seat of the self. When two individuals are in the presence of one another's timbres, durations, tones, breath, and spit, a third space is constituted within which their voices intertwine. The voice speaks here through embodiment, as it is the route of sonic information to the infrastructure of the body. It does not just carry words—it *places* the speaker and everything they feel, along with the words. A voice can carry no content other than its own fleshy encasement—of necessity, an intermediary force before it can signify.

This is a perspective shared in Paolo Virno's (2015) recently translated *When the Word Becomes Flesh* (2015). He offers an example of this "intermediation," or "in-betweenness," by examining the biological and historical intersections of language, voice, and speech as an intermediate zone that constitutes subjectivity, which he explains in these terms:

> The pure I, subject to the a priori categories organizing thought, most certainly is not a substance, neither is it an ineffable presupposition: it consists in a linguistic act. And a linguistic act can't but be extrinsic, phenomenal, perceptible. [...] The same is true for "I speak": its vocalization is an integral part of its meaning. The laryngeal motion that allows us to perform the locutory action we claim to be realizing is the sensible interface of the synthetic unit of apperception. The voice, which in the speech act "I speak" acquires the status of an indispensable logical requisite, reifies the transcendental subject each and every time. [...] The foundation of categorical thought is not a thought, but an action. An audible action. (2015, pp. 160–161)

Virno's (2015) intermediate zone is thus a stone's throw from LaBelle's (2014) theory of the mouth-voice. It is a call for understanding the voice as an intertwining of the living biological body with the multifaceted environments that the body gives voice to, while also reflecting the ways that environments return voice to the body as it responds to those environments—an example could be that the cessation of stage fright allows the speaker to once again claim their voice, previously frozen in fright. The speaker is a gifted virtuoso of the voice and of language, Virno argues, needing only the ability to speak and to use their voice, acting towards the rearrangements of their environments, and exploring the otherwise indeterminate relations between life and language. The articulation of voice and of language at once structure the speaker's environment; more importantly, these articulations announce the presence of the speaker as a sentient and communal being, made visible by their voice. The voice is therefore ritualistic, in that every speaking event instantiates an announce-ment at the place of the body, as well as the position of the subject in that place.

Speech, Virno (2015) argues, is always already tied to acts, to locution-ary events, to performances. The voice produces itself while indexing its production. He further posits that no performance matches the sound of the voice; in the act itself, the voice, its physiology and its breath, are the peak experience of an articulation in sound, wherein the voice produces a *flatus vocis*: to say that you are saying in the saying of what you say. Virno uses the example of the hail to illustrate this idea: "*Hey!*" he writes, is an example of how the "fact of speaking," the very embodied capacity to emit sound, is paramount to the meaning of "*Hey!*" And rather than ground the fact of speaking socially or culturally, he ascribes this fact to the body's existential presence, a living body that must breathe to voice. Thus, speaking is caused by and is the cause of the living body: as such, there is no separate entity *behind* speaking. Virno describes this condition as "absolute performative." That is, the generic faculty of language (its articulation in history and its representiational potential for communities and for collectives) is tied up with the signifying voice, a physiological capability, and a structuring agent: "The rhythm of the breath [is] the contractions of the diaphragm, the tongue pushing against the teeth (the physiological) represent at every time the power to speak" (Virno, 2015, p. 56).

Certainly, Virno's account admirably multiplies the voice's mediational potentials beyond its containment in the mouth, expanding its place to include flesh, motion, vibration, and intention; more elements in the voice's infrastructure of audibility. The voice is also confirmation of the "liveness" of the individual to which it belongs, even in the case of a shallow breath or a small twitch in the corner of the mouth. And while the voice speaks to how its speaker is situated in a specific and local spatiotemporal framework, it also speaks to subjectivity as a condition of being alive. The voice is simultaneously mediated, embodied, technical, and aesthetic.

To return to LaBelle (2014), it thus remains curious that he would focus so much attention on the mouth, or indeed any one organ, as though the voice's abstraction into any one of its "containers" helps explain it ecolog-ically. The danger in his abstraction, I think, perpetuates a hypothesis of the voice that has long been in circulation: that it is at once a part of the body and apart from the body. But Virno's problem is different: his voice floats free of the technological networks that have contributed towards the bifurcation that extend its continued abstraction.

A New Theory of Voice

Here, following the historical journey presented, I want to briefly sketch how the scientific discovery of another organ contributed to both the elevation of the voice—so that it becomes a quality secondary to the body's visual fixture—and to an ultimately individualistic and anthrocentric theorization of the voice. From the practice of laryngoscopy, the voice emerged in the condition of a scientific bifurcation, *away from the experience of the voice as an event and towards the voice as a set of techniques.* In the next chapter, I move towards an ecological model of the *voicescape* by considering how the voice is entangled in social and technical networks.

But for now, much of these considerations can also be located discursively through the networks concerned with the invention of the laryngoscope. I want to stress that the voice is entangled in a complex set of relations with the body, technology, and the social, and is often theorized as an intermediary, mediator, or transducer between internal and external worlds. It is a well established perspective that often underlies the "floating voices" that permeate techno-culture, from phonographs to iPhones. It also supports the "schizophonia" that was diagnosed by R. Murray Schafer (1993) as the split between sound and source, endemic to modern industrialism and communications technologies. Against such reductionism, I return now to Bernard Stiegler's (2004) concept of general organology as it offers a unique route to think through the conceptual obstacles raised by the voice. General organology has three attributes: organs, organizations, and technics. It is the last of this triad that provides us with the most insight on the voice as an imaginary organ.

Stiegler (1998, 2010) claims that physiological organs co-evolve with the technologies that enable their capture and facilitate their understanding. This means that organs are deeply entwined with the social organizations through which knowledge surfaces—in other words, as noetic organs that are "not only organic, not only organizational" (Stiegler, quoted in Wambacq, Ross, & Buseyne, 2016, p. 4). As much as Stiegler's work is sometimes misinterpreted as a fetish for the technic, the technic accounts for an emergence and intermingling of individuation within and across social, technical, and organic milieux. This emerging individuation arises from matter becoming something new. Such a process underlies the very conceptual matter of a general organology, less as a method than as an approach, which Stiegler defends as "a way not only of posing questions, but of letting oneself be put into question" (Stiegler, quoted in Wambacq et al., 2016, p. 4).

In the context of the voice studies turn, which alludes to the bifurcation between voice and body, the voice is taken along a rather unsurprising route of performance and aesthetics (Young, 2015). Dominant views on the subject tend to come either from pedagogists, musicologists, or performance specialists (see Eidsheim, 2015). The voice studies turn, detached from the overly theorized musings of Lacanian- or Derridean-informed philosophy, offers a new agency of the subject, which refutes much of the talk about interpellation endemic to theorizations of voice throughout the twentieth century. Contemporary voice studies scholarship looks towards the affective materiality as well as the techno-cultural contexts within which voice is situated and articulated.

Voice sciences can also work in ways that are less intended to repair voice estrangement than to disrupt the naturalness of voice embodiment. So, the estranged voice can be used as a form of resistance: tone, volume, intensity, resonance, and pitch are radically scrambled so as to defer any ideological attachment we might have to a voice. Such a tactic was famously used by American artist and performer Laurie Anderson, who regularly used prosthetic voice-altering technologies (such as the vocoder) either to multiply her voice into many voices (such as her piece "O Superman") or to lower the pitch of her voice to enter into a culturally ascribed masculine register ("Mach 20"). Anderson's vocal alterations might be thought of as a critique of the voice image, as she uses technology to evaluate the presupposed values that she embodies, at once a critical and creative gesture. Anderson thus pulls the voice out of the body and purposely displaces it to rupture the intersections of identity, subjectivity, and body.

Such vocal ruptures are central to Stiegler's general organology, which considers every technical and social situation through which a component of human experience (such as the voice) is or can be knowable. General organology allows us to think about the coming-into-being of individuated substances, like a "noetic organ" such as the voice, but this theory constitutes them as inseparable from the technical or social ties that are intrinsic to the process of becoming. One of Stiegler's recent examples is of brain-imaging research on children who play video games, which some claim condition the brain to synthesize the capacity for deep attention and hyperattention. Some even suggest that schools might do best to accommodate this accidental change in cognition rather than stigmatizing the development as a problem (Stiegler, 2010, p. 74). General organology centralizes the body itself as an evolving concept entwined in prosthetics, organic matter, and ideas, while stressing—in a Deleuze-inspired moment—that "prosthesis is not a mere extension of the human body; it is the constitution of this body qua human" (Stiegler, 1998, p. 152). This process he terms "epiphylogenesis":

> A recapitulating, dynamic, morphogenetic (phylogenetic) accumulation of individual experience (epi), designates the appearance of a new relation between the organism and its environment, which is also a new state of matter. If the individual is organic organized matter, then its relation to its environment (to matter in general, organic or inorganic), when it is a question of a who, is mediated by the organized but inorganic matter of the organon, the tool with its instructive role (its role qua instrument), the what. It is in this sense that the what invents the who just as much as it is invented by it.

Namely, the voice is an organ of reproduction, but it is also an organ of production—it is also not an organ at all, but an accumulation of organs (and technologies) that produce a capacity for speaking an individual or collective experience, for "giving account of oneself" (Butler's [1997] axiom). It is the uppermost projection of the body's dynamic capacity for volition in communication. The voice, to expand on Stiegler, is positioned as in-between internal and external, purely demarcating this potential through the organic corporeal and inorganic institutional forces that both enable such a capacity and restrict it. It reflects, especially if not exclusively, the capacity for volition—a quality that is becoming a central defining feature in voice studies.

Concluding Remarks

The purpose of this chapter has been to reconcile the voice as a dynamic site of transduction between embodiment and estrangement, located particularly within its medical and scientific discourses. It is indeed exceedingly difficult to talk about what the voice *is* since "it" is ongoing and continually being worked on. And, certainly, technological innovations fully intend to supply the voice with the affective material and immaterial nuances that demarcate the volition of human communication.

This discussion of voice imaging, of making the voice knowable, offers another way of thinking through the increasingly important role the voice plays in contemporary institutional and extra-institutional politics. With an expanded definition of voice—as an imaginary organ—we might be able to ask what it means to give voice to a contemporary social issue, which is itself a postulate of the voice that deserves its own historically grounded investigation. There will always be celebrities with the best voices, those singers who possess something we desire, but so long as we hold them in a superior position over their deviations, we run the risk of excluding a more democratic conceptualization of speaking and of communication. This is why Stiegler's (1998) general organology is so commendable: it causes us to redefine human bodies as entangled in the technics which facilitate the self-reflexivity of those bodies. It is for this reason that we must push further to explore the estranged configurations of the voice, to generate new frameworks for the voice's belonging to an expanded sense and sensation of embodiment.

The next chapter expands on this general organology by placing the voice in its social relations with other-than-humans.

References

Aronson, A. E., & Bless, D. M. (2011). *Clinical voice disorders* (4th ed.). New York, NY: Thieme Publishing.

Back, L. (2007). *The art of listening*. London, UK: Berg.

Bailey, B. (1996). Laryngoscopy and laryngoscopes—Who's first? The forefathers/four fathers of laryngology. *The Laryngoscope, 106*(8), 939–943.

Baron, B. C., & Dedo, H. H. (1980). Separation of the larynx and trachea for intractable aspiration. *The Laryngoscope, 90*(12), 1927–1932.

Barthes, R. (1977). The grain of the voice (S. Heath, Trans.). In R. Barthes (Ed.), *Image-music-text* (pp. 179–189). London, UK: Fontana.

Blackman, L. (2000). Ethics, embodiment and the voice-hearing experience. *Theory, Culture & Society, 17*(5), 55–74.

Blackman, L. (2010). Embodying affect: Voice-hearing, telepathy, suggestion and modelling the non-conscious. *Body & Society, 16*(1), 163–192.

Blackman, L. (2012). *Immaterial bodies: Affect, embodiment, mediation*. London, UK: Sage.

Blackman, L. (2016). The challenges of new biopsychosocialities: Hearing voices, trauma, epigenetics and mediated perception. *Sociological Review Monographs, 64*(1), 256–273.

Bozzini, P. (1807). *Der Lichtleiter oder Beschreibung einer einfachen Vorrichtung und ihrer Anwendung zur Erleuchtung innerer Höhlen und Zwischenräume des lebenden animalischen Körpers*. Landes-Industrie-Comptoir.

Butler, J. (1997). *Excitable speech: A politics of the performative*. London, UK: Routledge.

Cavarero, A. (2005). *For more than one voice: Toward a philosophy of vocal expression* (P. A. Kottman, Trans.). Stanford, CA: Stanford University Press.

Chatziprokopiou, M. (2015). Lamenting (with the) "others", "lamenting our failure to lament"? An auto-ethnographic account of the vocal expression of loss. In K. Thomaidis & B. Macpherson (Eds.), *Voice studies: Critical approaches to process, performance and experience* (pp. 149–161). London, UK: Routledge.

Connor, S. (2000). *Dumbstruck: A cultural history of ventriloquism*. Oxford, UK: Oxford University Press.

Conrad, P. (2007). *The medicalization of society: On the transformation of human conditions into treatable disorders*. Baltimore, MD: Johns Hopkins University Press.

Czermak, J. N. (1861). On the laryngoscope and its employment in physiology and medicine. *New Sydenham Society, 11*, 1–79.

Deleuze, G. (1986). *Cinema 1: The movement image.* London: A & C Black.

Deleuze, G., & Guattari, F. (1987). *A thousand plateaus: Capitalism and schizophrenia* (B. Massumi, Trans.). Minneapolis: University of Minnesota Press.

Di Matteo, P. (2015a). Capture of speech in (dis)embodied voices. In K. Thomaidis & B. Macpherson (Eds.), *Voice studies: Critical approaches to process, performance and experience* (pp. 104–119). London, UK: Routledge.

Di Matteo, P. (2015b). Performing the entre-deux: The capture of speech in (dis)embodied voices. In K. Thomaidis & B. Macpherson (Eds.), *Voice studies: Critical approaches to process, performance and experience* (pp. 90–103). London, UK: Routledge.

Dolar, M. (2006). *A voice and nothing more.* Cambridge: MIT Press.

Dyson, F. (2009). *Sounding new media: Immersion and embodiment in the arts and culture.* London, UK: University of California Press.

Echternach, M. (2016). Magnetic resonance imaging of the voice production system. In R. T. Sataloff (Ed.), *Professional voice: The science and art of clinical care* (4th ed.). San Diego, CA: Plural Publishing.

Eidsheim, N. S. (2015). *Sensing sound: Singing and listening as vibrational practice.* Durham, NC: Duke University Press.

Eidsheim, N. S., & Mazzei, L. A. (2019). *The Oxford handbook of voice studies.* Oxford, UK: Oxford University Press.

García, M. (1855). Observations on the human voice (Proceedings of learned societies). *Philosophical Magazine, 10*(65), 217–229.

García, M. (1881). On the invention of the laryngoscope. *Transactions of the international medical congress, seventh session* (pp. 197–199). London, UK: International Medical Congress.

Gvion, L. (2016). "If you ever saw an opera singer naked": The social construction of the singer's body. *European Journal of Cultural Studies, 19*(2), 150–169.

Helmholtz, H. L. F. (1885). *On the sensations of tone as a physiological basis for the theory of music* (A. J. Ellis, Trans.). London, UK: Longmans, Green.

James, P. (1885). *Laryngoscopy and rhinoscopy in the diagnosis and treatment of diseases of the throat and nose.* London, UK: Baillière, Tindall, and Cox.

Järviö, P. (2015). The singularity of experience in the voice studio: A dialogue with Michael Henry. In K. Thomaidis & B. Macpherson (Eds.), *Voice studies: Critical approaches to process, performance and experience* (pp. 25–37). London, UK: Routledge.

Kanngieser, A. (2012). A sonic geography of voice: Towards an affective politics. *Progress in Human Geography, 36*(3), 336–353.

Kuhnke, E. (2012). *Persuasion and influence for dummies.* Hoboken, NJ: Wiley.

LaBelle, B. (2014). *Lexicon of the mouth: Poetics and politics of voice and the oral imaginary*. London, UK: Bloomsbury.

Levitt, R. (2013). Silence speaks volumes: Counter-hegemonic silences, deafness, and alliance work. In S. Malhotra & A. E. Rowe (Eds.), *Silence, feminism, power: Reflections at the edges of sound* (pp. 67–83). London, UK: Palgrave Macmillan.

Ling, C., Li, Q., Brown, M. E., Kishimoto, Y., Toya, Y., Devine, E. E., & Welman, N. V. (2015). Bioengineered vocal fold mucosa for voice restoration. *Science Translational Medicine, 7*(314). Retrieved from https://stm.sciencemag.org/content/7/314/314ra187.

Mazzei, L. A. (2013). A voice without organs: Interviewing in posthumanist research. *International Journal of Qualitative Studies in Education, 26*(6), 732–740.

Mazzei, L. A., & Jackson, A. Y. (2012). Complicating voice in a refusal to "let participants speak for themselves". *Qualitative Inquiry, 18*(9), 745–751.

Mazzei, L. A., & McCoy, K. (2010). Thinking with Deleuze in qualitative research. *International Journal of Qualitative Studies in Education, 23*(5), 503–509.

Merkur.de. (2016). MRT-Aufnahmen von Michael Volle bei "Lied an den Abendstern" [Gruseliges video]. Retrieved from https://www.youtube.com/watch?v=GCluRCd2YuM.

Merleau-Ponty, M. (1962). *The phenomenology of perception* (C. Smith, Trans.). London, UK: Routledge & Kegan Paul.

Miller, D. (2006). *Be heard for the first time: The woman's guide to powerful speaking*. Sterling, MD: Capital Books.

Mrázek, J. (2015). A sea of honey: The speaking voice in Javanese shadow-puppet theatre. In K. Thomaidis & B. Macpherson (Eds.), *Voice studies: Critical approaches to process, performance and experience* (pp. 64–76). London, UK: Routledge.

Neumark, N., Gibson, R., & van Leeuwen, T. (2010). *Voice: Vocal aesthetics in digital arts and media*. Cambridge: MIT Press.

Pantalony, D. (2009). *Altered sensations: Rudolph Koenig's acoustical workshop in nineteenth-century Paris*. New York: Springer.

Peters, J. D. (2015). *The marvelous clouds: Toward a philosophy of elemental media*. Chicago: University of Chicago Press.

Pieters, B. M., Eindhoven, G. B., Acott, C., & van Zundert, A. A. J. (2015). Pioneers of laryngoscopy: Indirect, direct and video laryngoscopy. *Anaesthesia and Intensive Care, 43*(Suppl. 1), 4–11.

Rahaim, M. (2019). Object, person, machine, or what. In N. Eidsheim & K. Meizel (Eds.), *The Oxford handbook of voice studies* (pp. 19–34). Oxford: Oxford University Press.

Ruppaner, A. (1867). The practice of laryngoscopy and rhinoscopy. *New York Medical Journal: A Monthly Record of Medicine and the Collateral Sciences, 6* (1), 1–25.

Rush, J. (1827). *The philosophy of the human voice: Embracing its physiological history; together with a system of principles by which criticism in the art of elocution may be rendered intelligible, and instruction, definite and comprehensive to which is added a brief analysis of song and recitative.* Philadelphia, PA: Printed by J. Maxwell.

Saner, E. (2012). Voice lifts: Something to shout about. *The Guardian.* Retrieved from https://www.theguardian.com/lifeandstyle/2012/sep/23/voice-lift-vocal-cord-treatment.

Schafer, R. M. (1993). *The soundscape: Our sonic environment and the tuning of the world.* Rochester, NY: Inner Traditions – Bear & Company.

Scott, T. (2010). *Organization philosophy: Gehlen, Foucault, Deleuze.* London, UK: Palgrave Macmillan.

Seiler, E. (1884). *The voice in singing.* Philadelphia, PA: J. B. Lippincott & Co.

Sholl, R. (2015). "Stop it, I like it!" Embodiment, masochism, and listening for traumatic pleasure. In S. van Maas (Ed.), *Thresholds of listening: Sound, technics, space* (pp. 153–174). New York, NY: Fordham University Press.

Sterne, J. (2003). *The audible past: Cultural origins of sound reproduction.* Durham, NC: Duke University Press.

Stiegler, B. (1998). *Technics & time 1: The fault of epimetheus* (R. Beardsworth, Trans.). Stanford, USA: Stanford University Press.

Stiegler, B. (2004). *De la misère symbolique 1.* Paris, France: Galilée.

Stiegler, B. (2010). *Taking care of youth and the generations.* Stanford, CA: Stanford University Press.

Thomaidis, K., & Macpherson, B. (Eds.). (2015). *Voice studies: Critical approaches to process, performance and experience.* London, UK: Routledge.

Tiainen, M. (2013). Revisiting the voice in media and as medium: New materialist propositions. *NECSUS: European Journal of Media Studies, 2* (2), 383–406.

Vallee, M. (2017). The rhythm of echoes and echoes of violence. *Theory, Culture & Society, 34* (1), 97–114.

Virno, P. (2015). *When the word becomes flesh: Language and human nature.* Los Angeles: Semiotext(e).

Wambacq, J., Ross, D., & Buseyne, B. (2016). "We have to become the quasi-cause of nothing—of nihil": An interview with Bernard Stiegler.

Theory, Culture & Society (online first). Retrieved from https://doi.org/10.1177/0263276416651932.

Wells, W. A. (1946). Benjamin Guy Babington: Inventor of the laryngoscope. *The Laryngoscope, 56* (8), 443–454.

Windsor, T. (1863). On the discovery of the laryngoscope. *The British and Foreign Medicochirurgical Review, Or, Quarterly Journal of Practical Medicine and Surgery, 31*(January–April), 209–210.

Young, M. (2015). *Singing the body electric: The human voice and sound technology.* London, UK: Ashgate.

Younis, R. T., & Lazar, R. H. (2002). History and current practice of tonsillectomy. *The Laryngoscope, 112*(S100), 3–5.

3

Sounding Between Bodies

The Infrastructure of Audibility: The Voicescape

The last chapter pointed to the voice's infrastructure of audibility as the technologies through which an individual voice comes to pass through an individual body; in this chapter I use a more expanded notion of the infrastructure of audibility to place the voice between those technological manifestations, so that imaginary organs now inhabit the spaces between bodies, places, things, and worlds. This infrastructure of audibility traces the surfaces of cities, science, animals, art, nature, and science, as opposed to presenting a standard contextualization of the sociotechnical imaginary. The purpose is to decentralize the laryngealcentric perspective of the human voice in place of a concept of sounding that captivates a landscape, soundscape, or, more specifically, *a voicescape*.

The sociotechnical imaginary still features in the contemporary sounds of the human voice—however, it is embedded in a more nuanced network of sensory apparatuses that prioritize sensation over perception and affects over bodies. New computational media assesses those bodily emissions that otherwise evade our conscious awareness, bringing them into

M. Vallee, *Sounding Bodies Sounding Worlds*, Palgrave Studies in Sound, https://doi.org/10.1007/978-981-32-9327-4_3

consciousness. Thus, the paradox of twenty-first century media is that sensory media operate beyond consciousness to bring our attention to what underlies consciousness, as pointed out by Mark Hansen (2015). Our data doubleness resides in our flow before and after awareness. That is why contemporary mediation occurs in the domain of sensibility itself. This means we need to expand the whole span of the human/environment relation, so that the environment occupies the human as much as the human is in the environment.

To illustrate this opening statement, I turn briefly to voice professionals. Voice professionals run the very real risk of experiencing a voice impairment that, if left untreated, would be beyond repair, thus stripping them of their very livelihood. While teachers, for example, are encouraged to visit ear-nose-throat specialists regularly should they experience early symptoms of dysphonia such as voice hoarseness, scratchiness, or breathiness, they rarely do so because voice damage creeps in slowly, then leaves its mark suddenly—often when it is too late for medical intervention and a full recovery. The isolation that dysphonic teachers feel is palpable. Among teachers who suffer from dysphonia, the voice is not individual; it is categorically contextual and relational, embedded in a complex system.

Is it possible, some engineers are thus asking, to construct a smart-technology voice pathology detection system that alerts a patient at the preliminary signs of voice fatigue, before the symptoms of dysphonia begin to develop? While the final diagnosis of such a voice disorder as dysphonia can only be made by a physician, people are increasingly undergoing self-diagnosis using smart devices. Thus, an objective means of voice monitoring progressively shifts the power of diagnosis and treatment to a network of things that communicate detections of vocal irregularities.

We are familiar enough with the experience of being recorded for a brief period of time. Usually this is done with the intention of being listened to by human ears at some point in the future. But how would it feel to be listened to, carefully, by a digital device designed to detect signs of voice fatigue? In other words, how would it feel to be cared for, at a distance, by an infrastructure that is designed to receive our audibility? The issue with voice monitoring, particularly with vulnerable groups such as the elderly or those in the voice professions, is that voices need to be extracted, analyzed, categorized, and diagnosed in real time to make effective interventionist strategies possible. M. Shamim Hossain, Ghulam Muhammad, and Atif Alamri (2017) claim that such strategies would be facilitated if voice monitors were tied into the infrastructure of smart cities, through healthcare big data analytics and cloud computing in particular. This technology would combine a smartphone or recording device with electroglottographic wavegram analysis; while the former seems easy enough to understand, the latter is slightly more challenging to convey—but here goes:

An electroglottographic (EGG) wavegram analysis measures the oscillations of glottal cycles over time (a glottal cycle is the period of time it takes to produce such vocables as the vowels in a speech). The EGG visualizes the duration of the glottal cycles and determines whether, over time, the glottal cycles are shortening, quickening, or showing signs of irregularity, such as fatigue. Hossain et al. (2017) propose this step in voice pathology detection because the voice signal and the EGG are recorded simultaneously into a cloud for data processing and analysis. This makes it possible for a physician, should the voice detection system alert a patient, to determine the nature of the problem by choosing sound samples from the voice signal, along with their corresponding images from the EGG. Such a paradigm simultaneously visualizes the voice with the EGG, but it also "voices" visualization through the corresponding audio recordings. It becomes a cross-modal component of voicing in the smart city.

When the voice is measured in real time, the body is technically entangled in care. If one were to register with a smart healthcare service provider, the person would upload the vibrations of their own body as part of a data cloud in a linkage to stakeholders such as hospices and recovery centres; the patient would wear a necklace EGG device that would connect with their smartphone, itself equipped with a microphone that picks up the acoustic signal of their voice and a method for aligning the EGG with their voice data. The basic infrastructure combines three tiers of interaction (edge services/devices, the healthcare media cloud, and smart technologies) so as to provide residents in smart cities with functions and services (Hossain et al., 2017). The first of the three, edge devices and services, refers to the intertwining nature of the smart device with the particular stakeholder they are connected to: the smart device connects to selected smart city healthcare stakeholders such as research centres, hospices, or treatment centres, to automatically detect abnormalities and cross-examine against other smart devices.

What makes this model especially fresh and interesting is that healthcare professionals would thus be enabled not just with data representation (the EGG) but with the voice signal samples so that they can determine, subjectively, a route of diagnosis that could potentially assist a smart city resident in real time, over space, without visiting the physician. Given that the special feature of this particular voice pathology detection system is synthesizing EGG signals with acoustic voice signals, it opens the possibility for subjective diagnosis by relevant stakeholders. It relies on a Gaussian mixed model probabilistic calculator trained to detect differences between normal and pathological voices, which is performed in the healthcare cloud computing environment.

The voice makes sense in the smart city because, similar to the sketches of acoustic surveillance by Athanasius Kircher (1650/1970) that contribute to our current notion of ubiquitous listening and the contemporary "pan-acousticon," it generates some important insights about the way that bodies are becoming incorporated into data, clouds, information processing, and into new modes of governmentality through vibration. Where Kircher's pan-acousticon transmitted the disembodied voices for the purposes of monitoring, the new smart communication devices are intended to return information (especially any information of an abnormal nature) to the user via their own short-range communication devices, such as their laptop or their smartphone, should they exhibit any early signs of voice disorder (Hossain et al., 2017). It situates the sciences, the city, smart devices, and the idea of an exteriority of the body within the smart city and an exteriority of the smart city within the body.

This simple case presents a challenge to the laryngealcentric voice from Chapter 2, if only because the smart city voice is embedded in a technical urban industrial system that determines voice quality for particular professions and specializations. This chapter presents the *voicescape* as a relation between the body and the environment, in communities between its fleshy and its technical entanglements, and its political vulnerability. The topological folds of the body fuse the spatial, the temporal, the animal, and the technological on the cusp of communicability. If the infrastructure of audibility spans the voice outward between bodies, institutions, species, entities, and objects, it thus demands that we think of it as relational.

Empathic Vocality

A sociological approach accommodates another perspective on the voice as an imaginary organ: rather than merely opening the body, it opens the body *outward.* Likely as a result of its highly individuated status, the voice—as an object and as a process—has been a heteroclite topic within sociological perspectives of the body. When this topic does surface, the voice is often framed in terms of its performative capacity, with its potential as a *relational* technique of the body presumed. Here, I briefly explore early sociological approaches to the human voice as a structured and structuring agent of empathy in social relations, before moving onto a more relational account of the way voices find means of escape. Herbert Spencer (1891), in particular, in his non-laryngealcentric approach to the voice, argued that the voice is fundamentally attached to the body's capacity for gesture. He posited that all sounds arise from gestures that can be tied to their broader contexts, where they tend to be repeated over and through time. A body in pain or a body in the throes of ecstasy will emit vocalizations that, through the socialization process, we come to associate with pain or pleasure.

Spencer (1891) connected the biological, physical, and social processes through the claim that the voice is the final product of an interaction between mental and muscular "excitements" (p. 404). As a result, he was attentive to questions of how the voice represented the lived experiences of the body through its physiological dynamics—particularly in its loudness, timbre, pitch, intervals, and variation. Since the sound of the voice results from the voluntary and involuntary gestures of the body (such as contracting ligaments and the explosion of breath), Spencer argued that we merge the sounds of joy or pain with the lived experience of joy or pain. Turning towards music as a special reserve for emotional experiences, he ultimately suggested that music is a rationalized catharsis for intense vocalizations of extreme feelings, ones we are generally unaccustomed to or discouraged from expressing publicly. Examples here could be the guttural screams of a death metal group, or YouTube's autonomous-sensory-meridian-response (ASMR), in which "whisper vlogs" act as extreme vocalizations (see Andersen, 2015). Spencer concluded that the voice is foundational to what he considered a natural law of empathy:

> Each of us, from babyhood upwards, has been spontaneously making them, when under the various sensations and emotions by which they are produced. Having been conscious of each feeling at the same time that we heard ourselves make the consequent sound, we have acquired an established association of ideas between such sound and the feeling which caused it. When the like sound is made by another, we ascribe the like feeling to him [sic]; and by a further consequence we not only ascribe to him that feeling, but have a certain degree of it aroused in ourselves: for to become conscious of the feeling which another is experiencing, is to have that feeling awakened in our own consciousness, which is the same thing as experiencing the feeling. Thus these various modifications of voice become not only a language through which we understand the emotions of others, but also the means of exciting our sympathy with such emotions. (p. 410)

Spencer (1891), I suggest, is ultimately interested in the infrastructure underlying the meanings and relationships of vocalization, which garners insights into our capacity for empathy with one another. While laryngoscopy offered a technical-physiological basis for vocal production, the sociological perspective examines the bodily excitements that produce those vocalizations. Thus, sensation and emotion are the underlying basis for the sounds the body makes. Spencer theorizes that the voice's fusion to excitement confirms for us the notion that certain vocalizations come attached with particular excitations. Thus, vocalizations are not necessarily qualities of the body but underlie the very capacity for a body to be recognized as experiencing the world. To wit, Spencer's implication is strictly social, that we are, hypothetically at least, capable of experiencing the inner emotional world of another by recognizing the fusion of vocalization with sensation. And by extension, in hearing another's vocalization we experience that very state ourselves (or a likened version of it, at any rate).

Spencer's interests would lead him eventually towards a sociology of music, which continues today as a specialized subfield of sociological inquiry. John Shepherd (1991, 2015) has been an influential advocate for such a field, again situating much of his arguments with references to the human voice. Shepherd argues that the human voice, far from being an emotionally irrational or affectively pre-cognitive modality of communication, should be understood (or rather felt or heard) as a complex somatic pathway to the structured and structuring dimensions of human sociality—but he does this without grounding the voice in its scientific network. By extension, he observes that musical instruments are extensions of the human voice, an aspect which can be seen in musical systems around the world. And while he concludes that it is possible to analyze the voice (usually in its exceptional and performative musical inflections) as a structuring agent of social processes, he is short on suggesting a method. Regardless of his methodological shortcomings, his conclusion about the human voice is central to voice studies: that it is understood best in its concrete relations instead of its theoretical abstractions.

Relationality continues to be of especial interest in voice studies. Voice studies scholars are particularly interested in the audible and vibrational dimensions of embodied life, and it is here that we see an intersection between sound studies and the body. The voice can be regarded as a sonic force that emanates from bodies, vibrates between bodies, and expresses the affective states of bodies. To be sure, we might think of the voice as the intermediary tissue between individuals, or the medium through which a body announces its own position as well as the positions of others in everyday life. Indeed, many scholars have taken up this idea:

- Nina Eidsheim (2015) argues that the voice is the *primae facie* site of the material and the sensuous.
- Brandon LaBelle (2014), discussed in Chapter 2, writes that the voice is "a struggle to constitute the body" (p. 5).
- Alecia Y. Jackson and Lisa Mazzei have pointed to the voice as the incorporeal site of becoming and resistance (see in particular Jackson & Mazzei, 2011).

The consensus, if it can be parsed out of the massively diverse literature on the voice, is that the voice is still material even while intangible.

If a voice is bound to the space it produces, it seems inappropriate to refer to "a" voice, just as Konstantinos Thomaidis and Ben Macpherson (2015) deny the voice's definitive article, "the." Instead, a more generalizable "voicescape" may be tenable— a term that, with a little rotation, could be reimagined as "voice-escape," calling out playfully to the fact that the voice arises from, escapes from, the body. Here, I briefly conceive of a voicescape as a sonic texture, wherein the voice's affective presence exerts social control. This process unfolds in multiple ways, as the voice is embedded in physical processes, transmitted across time and space, and acts as a connective tissue between humans, nonhumans, and their ecologies (including their stakeholders). It is thus an expanding site for individual, social, and cultural transformation.

A voicescape finds a significant place in contexts of caring and is especially fine-tuned to intimacy and touch. The voice of a foetus's mother, for instance, carries vibrations that contribute towards the infant's healthy gestation. Elizabeth McLean (2016) writes that parents of prematurely born infants in neonatal units are encouraged to connect with their infants using their voices. It can be a tool to (1) build a sense of self-identity as a caring parent, (2) create an emotional catharsis, (3) foster a feeling of connection, and (4) cope with difficult moments. Cognitive psychologists refer to such a vocalized connection between mother and infant as a "communicative musicality" that facilitates a bond between them after the birth: the voice is transmitted as an immersive vibration to an enclosed, but separate being (see Malloch & Trevarthen, 2009). Immersing the infant in a voicescape disrupts the distance neonatal care units impose on mother/infant relations due to medical necessity. In this way, the voice traverses the technical biosocial milieu of neonatal care: it speaks to the impossibility of touch and offers a solution to it. The voice, then, breaches a distance that, in voicing, it admits. This "foundational voicescape," as it could be called, cultivates a sensorium between entities that are otherwise cut off from the biosocial milieu of nurturing.

Intersections between biological and social entities demonstrate that the voice is primarily an intersubjective, rather than an individual, component of human life. The voice is itself a vibrational body that connects corporeal bodies. Since our more contemporary understanding of voice is limited to human utterance, it is understandable that we usually attribute voices to individual actors. The voice is significant for its apparent motion and lodging between and beyond human bodies. Communicative musicality, by extension, might thus stand expansion into different social networks, not only between people, but between human and nonhuman bodies. The infrastructure of audibility is thus relational and ecological, and reconfigures the relations between bodies and their environments. As such, the infrastructure of audibility has many multiscalar openings, and the imaginary organ of the voice can be sounded across a range of vocal expressions, such as in the vocal folds of an orangutan.

The Voice Is an Opening

Within the vocal folds of Rocky, the orangutan at the Indianapolis Zoo, scientists have discovered that, under controlled conditions, the primate has the capacity to "voice," evidenced by its ability to (1) mimic grunts emitted by researchers, and (2) manipulate the tones and timbre of its own grunts (Lameira, Hardus, Mielke, Wich, & Shumaker, 2016). This capacity, which some neurologists have described as a "freedom from immediacy" (Shadlen & Gold, 2004, p. 1229), is central to the concept of volition: it entails the capacity to simultaneously anticipate one's own actions (that is, the intent of the moment) and to reflect on those actions (which is to claim agency over them), thus going beyond the needs of the particular situation (Haggard, 2008). Using an ethnoprimatological methodology, in which a researcher engages in playful communication with the researched subject (Malone et al., 2014), not only contributes to knowledge about our shared primate ancestry, but introduces a practical, shared space between the researcher and the researched that encourages biosociality (Meloni, 2014).

The discovery of the primate voice builds upon decades of study on primate communication in controlled environments, which largely held that these animals are incapable of voicing in a manner similar to humans (Bolhuis & Wynne, 2009). In the orangutan experiment, however, researchers would lead a game of "do as I do" vocalizations (called "wookies"), which helped to establish a common ground in human and nonhuman communication. Lameira et al. (2016) noted that "being able to socially learn new voiceless and voiced calls would have, thus, effectively set the evolution of an ancestral hominid articulatory system on a course towards a vocal system fundamentally similar to modern speech" (p. 2). The research laboratory in this case becomes a kind of "semiotic playground" (Upton, 2015), wherein a set of controlled conditions ("restraints") allow researchers to interpret a vocalization (the "field").

Having recorded 5660 audio hours of the researchers and Rocky grunting at one another, they used the open-access Raven sound analysis software to locate moments of mimicry. However, recounting this experimental anecdote is less about which vocalizer followed which earlier one. More interesting here is the fact that each grunt anticipated something: each vocalization, each sound, pushed against the received wisdom about primate capabilities such that new possibilities emerged, even as each orangutan grunt pursued the offered sustenance reward. Indeed, the sonic was not the object of study—rather, it was the medium through which a virtual archaeological site (the vocal folds of the orangutan) were opened up. Similarly, sound is not the object of study here: instead, sound *opens onto the horizon of a trans-species community*. As the capacities of the vocal folds increasingly intertwine, this horizon moves ever closer.

As I am interested in the voice as an *opening*, it seems most pressing to theorize this term, as what other word is there that so well captures the embarkment on new territories through sound? In keeping with Spencer's (1891) reading of sensation as the infrastructure of vocalization, I understand vocalization as the infrastructure of opening. Opening references the process of a negative space becoming an entity. Colloquially, we know well enough what an opening might be—it arrives more readily to our understanding than does *affect* or *becoming*, as we are well aware of what an opening is the moment we are in an opening. Even though affect or becoming are ideas that we ought to be more familiar with as describing experience, in fact, often they only make sense if they are explained. Openings are all around us, revealing potential relations all around us. Notably, opening does not imply the death knell term of "discovery." No scientist or technological innovator thinks that a sound technology is going to discover something that is already there, such as a shared ancestry. Instead, it opens the possibility for a new relation.

Opening is a non-representational way of thinking through difference. The term is so common that it would seem almost more appropriate and current to affix *affective* before it, like an "affective opening." But openings have intensive implications. For instance, for Gilles Deleuze, for whom thought was less a mental than a concrete force, an opening, as a thought, is transformational—it transforms the boundaries and edges between self, other, environment, and the technological mediations through which transformations take place. What, exactly, does an opening open onto? Staying with the example of the orangutan, opening is the infrastructure of becoming; every "becoming-with," "becoming-alongside," "becoming-in" is preceded by the infrastructure of desire to open onto new mediated sites—to make entities of relations.

Within the sound/voice studies context, opening also holds onto the underlying necessity for contemporary bioacoustics research to reach into and between the sounds of animals, to respect their individuated vibrations alongside their ecologies, habitats, and relations. Ecology and habitat are not idealized categories of nature here, but instead respect that the sound transmission captures a pure ecology and a pure habitat between the researcher and the researched. Opening also points to the way in which sounds themselves, such as the grunts of voice between entities, constitute a body of relations. Such relations set forth a rule of sounding in general (rather than merely being a unique case), in which sound, the sonic, and vibrations necessarily are persistently relational, grounded in an empathic vocality, which itself can be thought of as an infrastructure of audibility for a range of vocal expressions that span human and nonhuman vocalizations—a series of voicescapes.

Base Voicescapes

Is a yawn an embodied infrastructure of sociality? The range of vocalizations shared between animals and humans invites an account of the empathic envelopments of voice. For instance, we are familiar enough with the "contagious yawn" that arrives in the mid-afternoons of conferences and classrooms. The yawn is theorized as the kernel of evidence for our capacity for empathy. Primate species, for instance, are more likely to yawn in empathy with yawning kin than yawning strangers. Dogs will yawn under similar conditions, so long as the one they are imitating is known to them, and so long as they are genuinely yawning. The dog's ability to "catch" its human-companion's yawns is attributed thus to an underlying emotional and empathic relation that is based on the sound of the voice: they fuse the sound of their human-companion's voice with their face and their body. In humans, contagious yawning develops around the age of four, once children are capable of identifying the emotions of others; likewise, dogs are said to develop contagious yawning for specific individuals once they age past seven months. Thus, yawning is, for evolutionary biologists, a signal of mature attachments among animal group members. Extending this notion of contagion, the voice and the larynx has been replaced with the voice/mouth/lung assemblage. There is no voice, in other words, without the context under which the voice is articulated, through a yawn, a cough, a cry, or a laugh.

Laughing, much like yawning, is another base vocalization that is theorized as having evolved before speech, since the neural circuits connected to laughing are embedded in the most ancient parts of the brain. Also, like yawning, laughing is said to be shared across human and nonhuman species. Rats, for instance, emit ultrasonic laughter when they are tickled on the nape of the neck, linked to their subcortical brain, a region shared with humans. Evolutionary biologists also agree that laughter signals cooperation, since we all know that laughing is nearly impossible to fake. Vocal control research also suggests that subjects are easily able to detect the difference between volitional laughter and spontaneous laughter. It would be spurious to suggest that laughter is unaccompanied by other bodily gestures, such as the dog's involuntary exhalation (the "dog-laugh") combined with a play-bow. The dog-laugh is intended to open the space of play.

Grunts, coughs, laughs, yawns, and other sounds grant the voice its relational and inter-corporeal status. The bare fact of the voice is that it vibrates on many levels simultaneously and contains many sounds. For humans, the lungs inhale then quickly exhale through the vocal folds of the larynx, triggering their oscillation, which vibrates through the throat and into the sinuses, and are rounded into shapes ("formant" vibrations) by the muscles in the mouth. It is the production of these formants that evolutionary biologists hold as the particularly distinct human capacity to produce different phonemes. W. Tecumseh Fitch (2000), currently the leading authority on bioacoustics as it pertains to evolutionary biology, explains that the voice contains a massive amount of information. Indeed, a spectrogram analysis demonstrates that an animal's vocal production occupies a relatively small amount of space on a frequency chart, whereas a human voice occupies a practically full spectrum account. This is because the human voice resonates all at once between a source and a filter: there is the source, the breath, as it vibrates the vocal fold through the upper region of the body, and there is the filter, which is the production of formant frequencies. It would be overly simplistic to separate the source from the filter (designated as the larynx and the mouth), since it is the larynx's position which determines a greater range of formant frequencies; this refers in particular to a tongue that has more freedom to move both vertically and horizontally, and has a capacity to produce sound differences that are only microscopically distinguishable from one another.

Evolutionary biologists prefer formants to the more general descriptions of timbre, quality, or affect, because formant specifically references the physical muscular production of different sounds. Certainly, there is a tidy binary here between the unintended physiological production of sound (the inhalation and exhalation of air and its sympathetic vibrations) and the intentional physiological shaping of sound for the purposes of generating specific information. And while all species have the capacity to produce source/filter vocalizations, human beings produce far more specific and detailed sounds. To offer insight into the primacy of the filter (mouth) over the source (larynx), Fitch points to the act of whispering, which requires no source vibrations, only filter.

To return to the example of the orangutan, researchers are free to play since they do not need to listen for the subtleties of vocalization themselves during experiments. These data are captured using inexpensive recording technologies capable of storing large high-definition datasets with high-accuracy analytic capacities. The technologies are integral to bioacoustics, with the datasets at the heart of the discourse of conservationist interventions (Towsey, Wimmer, Williamson, & Roe, 2014). For example, when an animal group's collective calls begin to weaken, these changes are read as signs of the conditions of their environment. Michael Gallagher (2015) thus includes bioacoustics recording as a type of "nature style" field recording, since handheld recorders and autonomous recording units (ARUs) are non-intrusively positioned in environments where they pick up otherwise imperceptible vibrations, which can be visualized on spectrographic displays for scientific measurement.

These visualizations make apparent the many animal calls that occur below or above normal human hearing: mice emit ultrasonic mating calls (Portfors, 2007), plants emit infrasonic and ultrasonic vibrations when distressed (Gagliano, Mancuso, & Robert, 2012), and elephants communicate over great distances using infrasonic vibrations to locate migrating herds (Herbst et al., 2012). Whereas earlier research in bioacoustics was intended to identify animal species for the purposes of enumeration—as though their sonic emissions were the mechanical and physiological byproducts of an animal's presence—contemporary bioacoustics research reads animals' sounds as *messages*. These sounds are agents in the construction and maintenance of an animal's environment. The technologies involved in this research have migrated from instrumental (species identification) to being part of a more complex network and co-constitution of conservationist intervention, species protection, and the identification of shared ancestry.

The meanings of animal vocalization have occupied scientific discourse, used variously to prove diverse conclusions supporting or refuting animals' volition in the making of these sounds. René Descartes defended his notorious animal vivisections, for instance, claiming in court—against the complaints raised by neighbours who could not bear the screams bleeding into the streets any longer—that any of an animal's vocalizations were simply an automatic response, and that equating screams of pain with suffering was a category error. He asserted that animals were incapable of experiencing anything like an emotion or physical agony (see Bernstein, 1998). Such an exclusionary practice, involving the bifurcation of the body as mechanical automata distinct from its capacity for abstract speech, was also applied to human populations. As Jacques Rancière (2004) notes, Aristotle followed his own famous "man is a political animal" argument with a comparison to slaves as those who understand language, but do not possess it (pp. 4–5).

The infrastructure of audibility and the imaginary organ are not distinct from one another, but are instead folded into one another. It is entirely possible to think of the technological hardware that bring the voice to different orders of the social infrastructurally. But it is also possible to think of the voice's effects such as yawning, grunting, growling, and laughing, as components in the infrastructure of audibility of sociality. In the next section, I want to pursue the voice as an infrastructure of audibility for the becoming-animal of humans. This is to reposition the human in relation to a voice, not as a reflection of the self, but as a means of escape from identity through an open cavity of the body.

Opening the Voice and the Animal

Voice theorists are adept at theorizing the voice as an infrastructure of audibility. Douglas Kahn (1999), for instance, describes the voice as "the most widespread private act performed in public and the most common public act experienced within the comfortable confines of one's own body" (p. 7). It is often said that the voice passes through the most deeply personal folds of the self, but is simultaneously figured as a lagniappe of the social contract, as well as a functional requisite of social transformation. This bifurcation, however, is also pernicious. Voice studies, a subset of sound studies, derives from a tradition that takes the voice as the ultimate cue of estrangement from the body. This legacy infuses voice theories. They are so entirely steeped in the idea that modern subjectivity suffers from "voice estrangement" that this is the normative ethos. Following from this, the voice is reified as something once embodied, occasionally estranged, and eventually reunited with either the individual or collective, which the voice is obliged to represent.

As yet another turn on the voice as an opening, I turn now to Deleuze and Guattari's (1987) linguistic theory of incorporeal transformation as a segue into their hypothesis that the voice is *fundamentally* a relational site, summarized by their deceptively simple premise that there are always "all manner of voices in a voice" (p. 77). Deleuze and Guattari agree that the voice is central to the expression of the body, but resist talking of it as though it were "estranged"; instead, they conceptualize the voice as "far ahead of the face, very far ahead" (p. 333) and persistently between the "in-corporeal transformations" of bodies. These incorporeal transformations are especially characteristic of the voice, which they frame as the central conceptual apparatus of the voice. From this position, they convincingly theorize the voice as a site of possibility by precipitating change in place of representing subjectivity.

Deleuze and Guattari's (1987) theory of the voice begins with their unique approach to language. For them, language does not communicate, nor does it inform. Language activates incorporeal transformations—it moves from sound to sound, based on various units. They say of this that "[if language] always seems to presuppose itself, if we cannot assign it a non-linguistic point of departure, it is because language does not operate between something seen and something said, but always goes from saying to saying" (p. 85). Deleuze and Guattari argue for a pragmatic approach to language that is organized around two types of enunciation: *order-words*, or statements that "arrest," and *pass-words*, statements that "move." The ties to Deleuze's Spinozist leanings should be obvious here, especially the joyful/sadness affects that respectively excite and delay movement, a binary oscillation which Deleuze and Guattari describe like this:

> [T]he order-word is also something else, inseparably connected: it is like a warning cry or a message to flee. It would be oversimplifying to say that flight is a reaction against the order-word; rather, it is included in it, as its other face in a complex assemblage, its other component. (1987, p. 118)

They resist the nature of language as bound to signification and repre-sentation, and instead view it as entangled with capacities that order-words can transform, Thus, Deleuze and Guattari (1987) insist that the opera-tional unit of language is thereby pragmatically political. However, their reference to the political is not limited to authorities who administer order-words. They are not concerned with interpellation, but rather with how the pass-word moves order-words towards new arrangements. They thus demand that order-words be understood as inherently pragmatic, since the enunciation can only occur with the physical apparatus of the vocal cords.

For Deleuze and Guattari (1987), the order-word is thus responsible for more than the performative or illocutionary act; it is the underlying organizational principle of all language. A statement is actual, having a real effect on bodily responses. This archaeological sense of "within-ness" demarcates how an "indirect discourse [is] the presence of a reported statement within the reporting statement, the presence of an order-word within the word" (p. 93). In particular, they argue that

> indirect discourse in no way supposes direct discourse; rather, the latter is extracted from the former, to the extent that the operations of significance and proceedings of subjectification in an assemblage are distributed, attributed, and assigned, or that the variables of the assemblage enter into constant relations, however temporarily. Direct discourse is a detached fragment of a mass and is born of the dismemberment of the collective assemblage; but the collective assemblage is always like the murmur from which I take my proper name, the constellation of voices, concordant or not, from which I draw my voice. (1987, p. 93)

Deleuze and Guattari (1987) emphasize the voice's role within the incorporeal transformation of indirect discourse. Indirect discourse is a manner of speaking that underlies all discourse and has a "real but not actual" effect between the bodies involved in that exchange, expressed in the "becoming-animal" of human subjectivity, which means leaving the humanist conception of the human behind in favour of an image of human as governed by affects, forces, and flows. The voice is the bellwether of transformation in its animalistic grunts, howls, and screams that are subdued and excused in everyday social interactions. These characteristics have implications for their philosophy of becoming, which is distinct from incorporeal transformation insofar as the becoming is the effect of having become transformed through indirect discourse. The "becoming-animal" of voice is an *opening continuum* between human and animal subjects, a voice that contains the identity of its possessor, but a voice that is made of the meat and tissue capable of adjusting the very size and presence of that body. With animals, for instance, their voice may be a warning cry, or be used to amplify the animal or diminish its presence: mating mice emit ultrasonic vocalizations, while dairy cows utter individual vocalizations that are distinctly tailored to their calves.

The body's many possible configurations are reflected in the voice; the voice therefore instantiates a time horizon of the multiple speeds and tempos through which the granular and hoarse affective distributions of the body (coughs, spasms, hiccups, growls, etc. [all signs of the voice's weakness from Chapter 2]) are distributed, multiplied, and come into being. The possibility for voice is located less in the loss of logos, than it is in the continuum between human and animal, where subjectivity may be located and represented. For instance, a scream or a growl may be the "intensity" of the human body escaping through the voice, phenomena that are a part of, but apart from, the body. For Deleuze and Guattari (1987), however, the voice is always already ahead of the body, the voice displacing the song and disrupting the unified body, comparable to the way an insect's scratch replaces the songbird's refrain "with its much more molecular vibrations, chirring, rustling, buzzing, clicking, scratching, and scraping. Birds are vocal, but insects are instrumental: drums and violins, guitars and cymbals" (1987, p. 308). In the spectrum of the voice, Deleuze and Guattari promote a more political ontology of voice that is based on the concept of variation and a "generalized chromaticism" (pp. 95–98).

Contrary to the laryngealcentric voice from Chapter 2, Deleuze and Guattari's (1987) generalized chromaticism celebrates so-called vocal aberrations as the capacity to produce continuous variation. Simply, generalized chromaticism is "a question of a highly complex and elaborate material making audible nonsonorous forces" (p. 95) such as noise. Within the realm of music, Deleuze and Guattari point to a rational chromaticism that adheres to a tonal centre (such as the music of the Western classical tradition, and Wagner's stretching of the dominant chord throughout the entirety of Wagner's *Tristan and Isolde* before its resolution). Here, chromaticism is pulled into the logic of form, such as the standardized sonata form. In contrast, generalized chromaticism applies more directly to the postserialist music of Alban Berg as well as the aleatoric performances of John Cage and the New York school. Such work opens the concept of sounding onto the noises we otherwise conceal, just as all attendees who shuffle nervously in their seats to Cage's 4'33", the famous silent piece, become the voice of the composition.

More closely attuned to voice, Danger Music #17 by Dick Higgins (1962), which instructs the singer to "Scream! Scream! Scream! Scream!

Scream! Scream!", speaks more appropriately to the idea that the voice, as Deleuze and Guattari write, may be "truly 'machined' [in that] it belongs to a musical machine that prolongs or superposes on a single plane parts that are spoken, sung, achieved by special effects, instrumental, or perhaps electronically generated" (1987, p. 107). Their theory of voice suggests that a subject will be incorporeally transformed when voice deterritorializes the language that holds the subject in its place. For a voice to "sound," it must break with the order of the non-sonorous, including the scratches of an insect and the rasp of the scream, giving up more of the oft-celebrated "grain" of the voice in favour of the grind.

In the opening pages of *A Thousand Plateaus*, Deleuze and Guattari (1987) ask plainly, "Is it not through the voice that one becomes animal?" (p. 5). This proposition is supported by the fiction of "Professor Challenger," an Arthur Conan Doyle character who is invoked in *A Thousand Plateaus* to stand for the ontology involved in deterritorialization and the timbres of voice and speech: "His voice had become hoarser, broken, occasionally by an apish cough [until] his voice had become unbearably shrill" (p. 80). Deleuze and Guattari continue on this analysis:

> Challenger was finishing up. His voice had become unbearably shrill. He was suffocating. His hands were becoming elongated pincers that had become incapable of grasping anything but could still vaguely point to things. Some kind of matter seemed to be pouring out from the double mask, the two heads; it was impossible to tell whether it was getting thicker or more watery. Some of the audience had returned, but only shadows and prowlers. "You hear that? It's an animal's voice."

Such a metamorphic dynamic is reflected in the artist Olivier de Sagazan's many performances of *Transfiguration*, which consists of him rubbing and disfiguring layers upon layers of wet clay over his head, sending him gasping and obsessively moulding the dripping contours of his face and hair, smashing his head with his fists, stabbing it with objects, and uttering sloppy pre-verbal cries with a fiery despair. In the various performances of *Transfiguration*, there is no voice in the body, but instead a violent eruption of the voice under the pressure of the head and face. In one moment, by slashing the clay with a razor over his mouth, releasing a desperate gasp that fills the performance space, a particularly ontogenetic moment has been articulated through the voice: the voice as the body's site of relation outside itself.

The example of Joan La Barbara, an American experimental vocalist and composer, further explicates the voice as an open site of possibility and transformation. Her voice, which she describes as the "original musical instrument," does not embody her subjectivity, her identity, or her positionality in the sense that her voice gives presence to her self. Neither does her voice sound like anything estranged. It is neither private nor public, if only because her voice sounds of anonymous, sovereign, self-reproducing matter. There is no statement, no speech, and she does not "give voice" to any pressing contemporary social issue. There is simply nothing to hear behind her voice; instead, she plays with the infinitesimal dynamics of *the* voice, not *her* voice, crossing thresholds of audibility, at once protracted and contracted, exploring the range of sounds produced by way of the vibrations between her diaphragm, lungs, trachea, jaw, buccal cavity, teeth, tongue, lip, nostrils, nasal cavity, hard palate, velum, pharyngeal cavity, larynx, and oesophagus. Her voice is not a medium but an ongoing production, expanding the conventions derived from muscle tissue that underlie song and speech, taking flight along a continuous variation in vocal production. If common knowledge raises the voice as the sonic emission of the body in its communication with others, Joan La Barbara breaks radically from this conception of the voice by way of a disidentification: To exploit the variations in vocal production, stretching the poles between tone and speech, and reformulating the very foundation of what it means to have a voice. Such continuous variations are evident in her early works, such as *Circular Song*, *Vocal Extensions*, and *Twelvesong*.

In her composition, *Twelvesong*, for instance, recorded in 1977 in a multitrack studio, La Barbara performs foundational tracks, which she describes as "steady tones, circularly sung on inhale and exhale to create a constant sound, micro-tonally separated from each other to create subtle beats" with "flutters (ululation), inhaled glottal clicks, gentle sighing glissandi, and birdlike sounds" (La Barbara, 2002, p. 38). Lasting 12 minutes and 12 seconds, and using a variety of signal processing and postproduction mixing, here she explores her technique of "sound painting": layering blocks of sound, as though layers of paint, that draw the listener into the voice's timbre.

While La Barbara crosses a threshold with her voice by exploiting its pre-verbal playfulness, or the voice's proprioceptive sensibilities, John Cage brings those sounds from the forms of speech within which they are embedded: to turn everyday situations into musical compositions, and in particular to attend to the unintended sounds of everyday life that become the event or the performance. One which makes particular resource of the voice is *Empty Words*, which effectively empties words of their order.

Empty Words was recorded in studio in California and live in Milan for 3000 students. In it, Cage reads passages of Henry Thoreau, eliminating parts of words and resting long on vowels, stretching and iterating them. The crowd, however, becomes restless, and what in the beginning of the recorded track is silence, turns quickly into loud boos, claps, and requests for rock songs. Cage describes: "The audience began making sounds after 15 minutes + continued with fireworks, dirt thrown, water, etc., dancing, singing, percussion. I also kept on going" (Cage, 2016, p. 477). Such vocalizations and protests do not disturb the performance inasmuch as they make the performance. Voice makes place in no systematic way, but a continuous variation, a terrifying event, unpredictable, and messy on all angles. The voice does not contain but sets free.

Voice artists thus reverse the flow between the infrastructure of audibility and the imaginary organ. They use the imaginary organ to elucidate the processes of which we are not necessarily conscious in our usages of the voice. They extract the bodily and embodied, but in such a way that renders the voice a communicative and relational agent between entities. If in the sociotechnical imaginary of the voice's historical emergence, we witnessed the body as it is produced through its voice and vocalization, we have in voice art the total disassemblage of the body through the imaginary organ, which destroys as much as it creates. In its relation, then, the voice is an opening towards the ethical, and one which places a newfound certainty in turning our attention to the voices of our environment, and what those voices are telling us about the way the world is on a course of continuous variation.

Conclusion: A Vibrational Ethics

To claim that voices are connected with artificial and networked entities, both material and immaterial, misses the mark for how the body, as a sociological category of experience, is projected by a series of extensions of movement, vibration, breath, and impulses. Erin Manning (2014) has pointed towards the idea that vibration precedes the formation of a palpable body:

> A body never pre-exists its movement. Total movement courses through all incipient form-takings (the edging into itself of "object," the shading into itself of "figure"). What actualizes as this or that displacement, this or that form, is but a brief instantiation of what the movement has become. (p. 165)

Movement resists rational containment and representation but can be transmitted and responded to as a spatial extension of the body. The examples from the voice highlight it as a locus of connectivity, which includes connecting with an environment. Sha Xin Wei (2013) writes about this kind of embodied catenation in immersive environments:

> [T]he strategy is to suspend or bracket certain conventions about what constitutes body, subject, or ego while trying to develop a working understanding of embodiment and subjectivation—the formation of subjective experience. Movement, and in particular gesture, is an arguably essential aspect of engendering human experience. But rather than taking "the body" or "cognition" for granted as conceptual starting points, I attend to the substrate matter in which gesture takes place—hence the interest in responsive and in particular computational media created for sustaining experientially rich, improvisational activity. (p. vii)

For Sha, gestures continue to make up a human experience, just as they did for Spencer. But gestures alone do not suffice; they are not autonomous but are instead ecologically entangled. Gestures do not fill a space; in defining or shaping a space, they spatialize. These gestures can move beyond humans to vibrate between and within the domains of things.

Such developments, belonging at least in part to the affective turn, have fructified new understandings of the body as immaterial and fluid, and have spawned extensive studies devoted to vibration (Henriques, 2010), rhythm (Crespi, 2014; Ikoniadou, 2014), sound (Oliveros, 2011), and movement (Manning, 2014). Studies such as these signal the massive impact that the affective turn has had on our understandings of the body. Works belonging (peripherally or directly) to this affective turn (see, for example, Clough, 2009, 2010; Glass, 2017; Massumi, 2002; Pedwell, 2014) have been instrumental in shifting the conceptual vantage point: bodies that have been conceptualized as visual things are becoming bodies as virtual entities. In other words, once concrete sites of social control, bodies are now real but not actual assemblages of desire, movement, flow, and force, which are engaged in communities that transcend human/nonhuman partitions.

Vibration is thus relational—and this relationality is what we are trying to similarly understand about the voice—for which the recent writing of Maria de la Bellacasa (2017) is especially valuable. She remarks on how care resonates between subjects as a way of generating new ethical and concrete entanglements. She argues that when we express care for another, we reach towards the discomfort or pain of another. Behind care is a very pragmatic and concrete intention. To care does not mean to solve or to cure; rather, caring opens the possibility for relationality. de la Bellacasa writes that "I care" is the admission that one cannot necessarily help another; instead, this gesture opens onto a space of empathy. It could lead to the possibility of helping, but this opening is an admission of the limits of care since it is only on the cusp of help, not actual help. To care, de la Bellacasa continues, forces one into concrete conditions, specific problems that resist abstraction: affective, ethical, and practical embodiments. In this way, she reclaims the care as situated. It is, echoing Maurice Merleau-Ponty's (1968) ontology of the flesh, an ethical relation of touching and being touched:

> [T]ouch's unique quality of reversibility, that is, the fact of being touched by what we touch, puts the question of reciprocity at the heart of thinking and living with care. What's more, the reciprocity of care is rarely bilateral, the living web of care is not maintained by individuals giving and receiving

back again but by a collective disseminated force. Thus conceived, the complexity of the circulation of care feels even more all-pervasive when we think of how it is sustained in more than human worlds. Care is a force distributed across a multiplicity of agencies and materials and supports our worlds as a thick mesh of relational obligation. (de la Bellacasa, 2017, p. 20)

The voice and voice disorder monitoring, to refer to the opening of this chapter, then, is touching at a distance, a posthaptic touch. It follows that de la Bellacasa's model helps us move beyond the idealization of the voice as belonging to any one body, towards voice as the propagational force between bodies.

With the development of bioacoustics and the extension of the voice into an ecology of relations, I claim here that the voice could be revoiced as vibrational, both inside and beyond the experience of subjectivity. It is the body's capacity for resonance, for vibration, for movement that helps us understand how the body is entwined in a network and how that network is elucidated and maintained through voices and voicings. Thus, in place of thinking about the voice as a reflection of subjectivity or an extension of the body, the "vibratology" that is proposed here becomes the very medium that makes relationality and affect possible. What this means is that vibration makes a body visible while also acknowledging its intangibility; the crucial element is neither voice nor body alone, but the dynamic interface between them as an intermediary tissue between bodies and their material entanglements.

Another more intimate aspect of shared voice lies in bioengineered vocal fold mucosa used for voice restoration, which involves the bioengineering of vocal fold fibroblasts and epithelial cells grown in organotypic conditions in a laboratory, and tested (1) in vitro, (2) in canine ex vivo (a dead dog's throat), and (3) in vivo in a humanized mouse, whose human adaptive immune system has tolerated the transplantation (Ling et al., 2015). The procedure, which promises a full airway, vibratory-to-acoustic output as demonstrated both in vivo and ex vivo, is also less likely to be rejected by the body since the cells from which vocal folds are engineered could be drawn from the patient's already damaged folds (2015). In real terms, the patient's voice has a new potential, one that retains the signifying potential of the voice, while also making its return to the body that nurtured its dissolution. Bioengineered vocal fold mucosa, in its calculated estrangement, reunites people with their own potential and their own capacity to "give voice." But it equally represents a new voice intersection between organisms and technology, and human and nonhuman communications.

Established, well-known, and well-trodden cultural theories of voice and vocalization tend to stop at human utterances. However, Lameira et al. (2016) have recently demonstrated that an entire human/nonhuman communication system is lost when we become awash in anthropocentrism. In particular, the field of bioacoustics research has opened up new ways of thinking through human/nonhuman relations and has revealed possible evolutionary connections to vocalization in primates. Notably, volition is at the centre of the discovery. With the advent of high-definition bioacoustics recorders and spectrographic analysis, insights into the voice have developed to the point that we can now drawing linkages between human and primate vocalization. While previous studies mostly failed to find such linkages, the orangutan's utterances of "wookies" has lent credible evidence to the hypothesis that nonhuman primates have the potential to volitionally produce grunts under tightly controlled circumstances. This demonstrates that they exhibit "real-time, dynamic and interactive vocal fold control beyond the species-specific repertoire" (Lameira et al., 2016, p. 2), which constitutes the capacity to signify deep laryngeal control. However, in opposition to such depth, the next chapter explores the ecological implications of thinking the voice beyond the body, as it traverses the surface of the earth in environmental monitoring practices.

References

Andersen, J. (2015). Now you've got the shiveries: Affect, intimacy, and the ASMR whisper community. *Television & New Media, 16*(8), 683–700.

Bernstein, M. H. (1998). *On moral considerability: An essay on who morally matters.* Oxford, UK: Oxford University Press.

Bolhuis, J. J., & Wynne, C. D. L. (2009). Can evolution explain how minds work? *Nature, 458,* 832–833.

Cage, J. (2016). To Mario Cavista. In L. Kuhn (Ed.), *The selected letters of John Cage* (p. 477). Middletown, CT: Wesleyan University Press.

Clough, P. (2009). The new empiricism: Affect and sociological method. *European Journal of Social Theory, 12*(1), 43–61.

Clough, P. (2010). Afterword: The future of affect studies. *Body & Society, 16*(1), 222–230.

Crespi, P. (2014). Rhythmanalysis in gymnastics and dance: Rudolf Bode and Rudolf Laban. *Body & Society, 20*(3&4), 30–50.

de la Bellacasa, M. P. (2017). *Matters of care: Speculative ethics in more than human worlds*. Minneapolis: University of Minnesota Press.

Deleuze, G., & Guattari, F. (1987). *A thousand plateaus: Capitalism and schizophrenia* (B. Massumi, Trans.). Minneapolis: University of Minnesota Press.

Eidsheim, N. S. (2015). *Sensing sound: Singing and listening as vibrational practice*. Durham, NC: Duke University Press.

Fitch, W. T. (2000). The evolution of speech: A comparative review. *Trends in Cognitive Sciences, 4*(7), 258–267.

Gagliano, M., Mancuso, S., & Robert, D. (2012). Towards understanding plant bioacoustics. *Trends in Plant Science, 17*(6), 323–325.

Gallagher, M. (2015). Field recording and the sounding of spaces. *Environment and Planning D: Society and Space, 33*(3), 560–576.

Glass, K. (2017). *Politics and affect in black women's fiction*. London: Lexington Books.

Haggard, P. (2008). Human volition: Towards a neuroscience of will. *Nature Reviews Neuroscience, 9,* 934–946.

Hansen, M. (2015). *Feed-forward: On the future of twenty-first media*. Chicago: University of Chicago Press.

Henriques, J. (2010). The vibrations of affect and their propagation on a night out on Kingston's dancehall scene. *Body & Society, 16*(1), 57–89.

Herbst, C. T., Stoeger, A. S., Frey, R., Lohscheller, J., Titze, I. R., Gumpenberger, M., & Fitch, W. T. (2012). How low can you go? Physical production mechanism of elephant infrasonic vocalizations. *Science, 337*(6094), 595–599.

Hossain, M. S., Muhammad, G., & Alamri, A. (2017). Smart healthcare monitoring: A voice pathology detection paradigm for smart cities. *Multimedia Systems*. Retrieved from https://doi.org/10.1007/s00530-017-0561-x.

Ikoniadou, E. (2014). Abstract time and affective perception in the sonic work of art. *Body & Society, 20*(3&4), 140–161.

Jackson, A. Y., & Mazzei, L. (2011). *Thinking with theory in qualitative research: Viewing data across multiple perspectives*. London: Routledge.

Kahn, D. (1999). *Noise, water, meat: A history of sound in the arts*. Cambridge: MIT Press.

Kircher, A. (1650/1970). *Musurgia universalis*. Hildesheim and New York: G. Olms (Reprint of the Rome, 1650 edition).

La Barbara, J. (2002). Voice is the original instrument. *Contemporary Music Review, 21*(1), 35–48.

LaBelle, B. (2014). *Lexicon of the mouth: Poetics and politics of voice and the oral imaginary*. London, UK: Bloomsbury.

Lameira, A. R., Hardus, M. E., Mielke, A., Wich, S. A., & Shumaker, R. W. (2016). Vocal fold control beyond the species-specific repertoire in an orangutan. *Scientific Reports, 6*. Retrieved from https://www.nature.com/articles/srep30315.

Ling, C., Li, Q., Brown, M. E., Kishimoto, Y., Toya, Y., Devine, E. E., & Welman, N. V. (2015). Bioengineered vocal fold mucosa for voice restoration. *Science Translational Medicine, 7*(314). Retrieved from https://stm.sciencemag.org/content/7/314/314ra187.

Malloch, S. E., & Trevarthen, C. E. (Eds.). (2009). *Communicative musicality: Exploring the basis of human companionship*. Oxford, UK: Oxford University Press.

Malone, N., Wade, A. H., Fuentes, A., Riley, E. P., Remis, M., & Robinson, C. J. (2014). Ethnoprimatology: Critical interdisciplinarity and multispecies approaches in anthropology. *Critique of Anthropology, 34*(1), 8–29.

Manning, E. (2014). Wondering the world directly—Or, how movement outruns the subject. *Body & Society, 20*(3–4), 162–188.

Massumi, B. (2002). *Parables for the virtual: Movement, affect, sensation*. Durham: Duke University Press.

McLean, E. (2016). Exploring parents' experiences and perceptions of singing and using their voice with their baby in a neonatal unit: An interpretative phenomenological analysis. *Article in Nordic Journal of Music Therapy, 11*, 1–42.

Meloni, M. (2014). How biology became social, and what it means for social theory. *The Sociological Review, 62*, 593–614.

Merleau-Ponty, M. (1968). *The visible and the invisible: Followed by working notes*. Evanston, IL: Northwestern University Press.

Oliveros, P. (2011). Auralizing in the sonosphere: A vocabulary for inner sound and sounding. *Journal of Visual Culture, 10*(2), 162–168.

Pedwell, C. (2014). *Affective relations: The transnational politics of empathy*. London, UK: Palgrave Macmillan.

Portfors, C. V. (2007). Types and functions of ultrasonic vocalizations in laboratory rats and mice. *Journal of the American Association for Laboratory Animal Science, 46*(1), 28–34.

Rancière, J. (2004). Introducing disagreement. *Angelaki, 9*(3), 3–9.

Sha, X. W. (2013). *Poiesis and enchantment in topological matter.* Cambridge: MIT Press.

Shadlen, M. N., & Gold, J. L. (2004). The neurophysiology of decision making as a window on cognition. In M. S. Gazzaniga (Ed.), *The cognitive neurosciences III* (pp. 1229–1243). Cambridge: MIT Press.

Shepherd, J. (1991). *Music as social text.* London, UK: Polity.

Shepherd, J. (2015). Music, the body, and signifying practices. In K. Devine & J. Shepherd (Eds.), *The Routledge reader on the sociology of music* (pp. 87–96). London, UK: Routledge.

Spencer, H. (1891). The origin and function of music. In *Essays: Scientific, political, and speculative, 2.* London: Williams and Norgate.

Thomaidis, K., & Macpherson, B. (Eds.). (2015). *Voice studies: Critical approaches to process, performance and experience.* London, UK: Routledge.

Towsey, M., Wimmer, J., Williamson, I., & Roe, P. (2014). The use of acoustic indices to determine avian species richness in audio-recordings of the environment. *Ecological Informatics, 21,* 110–119.

Upton, B. (2015). *The aesthetic of play.* Cambridge: MIT Press.

4

Sounding Ecologies

Transacoustic Communities

About 100 kilometres north of the Alberta Oil Sands, hidden amidst the burnt and twisted shards of lumber netted over freshly growing grass and sprouting pine, a common nighthawk nests on the ground, invisible to our eyes despite our attempts. Its own speckled pattern mingling with the black, ash, and tan of the land, the nighthawk sits unseen and unheard until one of the biologists whispers to the rest of us, "got it." I can see it: its form emerges from its surroundings under my own eyes like an image that surfaces through a magic eye test. It is an organism whose home only a year ago was engulfed in a 700,000-hectare forest fire. It lays there like a taxidermy prop, but breathing rapidly, seemingly unaware that we see it, its solid black marble eyes glistening with the silence of a life filled with waiting. It seems designed for this terrain, and even as I see it, I cannot comfortably say that I can fix it under my gaze—my vision cannot hold it in all certainty, which the nighthawk uses to its advantage when it explodes from the ground where it nests. When it senses that its young are under threat, the common nighthawk will produce a loud "wing-clap" as it arks dramatically through the air away from its makeshift nest, since they do not nest in trees. This one lands in front of us, clumsily, with its

© The Author(s) 2020
M. Vallee, *Sounding Bodies Sounding Worlds,* Palgrave Studies in Sound,
https://doi.org/10.1007/978-981-32-9327-4_4

plumage puffed out and its wing bent backwards, metres from its nest, in an attempt to distract us from its young; one of the researchers slowly positions his iPhone overtop two chicks left on the ground and takes a picture, after which we leave hastily to let the mother return to her pair of offspring.

The common nighthawk is difficult to sight: it is nocturnal, it blends in with its environment, it is notoriously elusive, and it is relatively quiet save for a nasally *peent!* in flight, and an explosive sonic wing-clap as it dives (Viel, 2014). Sighting nighthawks is painstaking work, especially while they are nesting, which is why biologists are turning increasingly to bioacoustics technologies for the purposes of identification and location exercises: an animal's sonic emissions serve as a reliable route of access to their location and their patterns of behaviour (Laiolo, 2010). What to do with these sonic emissions and how they play into the scientific imagination is the focus of this chapter.

In the context of the biological sciences, researchers who use bioacoustics are interested in animals' sounds in their ecological contexts, reading those sounds for what they might indicate about the security of biodiversity and the related concerns of ecological depletion (this context-based programme of research is what some call "ecoacoustics" [see Sueur & Farina, 2015]). Bioacoustics researchers use a variety of sound equipment to gather and analyze data: durable autonomous recording units (ARUs) can store years of information from within one location (Hutto & Stutzman, 2009), and backpack microphones strapped to animals' backs will track their sonic patterns as they move through space (Gill et al., 2016). Data are uploaded onto "listeners" that align the sounds with their appropriate species and such results are uploaded for international research centres and for international research teams.

I call this digital community of nighthawk researchers a transacoustic community, which expands on Barry Truax's (2001) definition of the acoustic community. He defines an *acoustic* community as an "information rich" system that uses "acoustic cues and signals [which play a] significant role in defining the community spatially, temporally in terms of daily and season cycles, as well as socially and culturally in terms of shared activities, rituals, and dominant institutions" (p. 66). But here, a transacoustic community *transcends* the immediacy of place, *transgresses* the boundaries of immediate community, *transfers* data into international research centres, *transduces* the visual into auditory analysis that has a better and higher definition, and *transposes* the audible into the visible. Because the sharing of acoustic data allows signs of population depletion and biodiversity loss to emerge, researchers' imaginations become a scientific tool for intervening in avoidable and undesirable futures. Thus, by being interested in how researchers are implicated in the infrastructures they spontaneously design, I work towards inverting that infrastructure. Indeed, I will argue that such encounters are almost entirely reliant on a specific form of imagination in which the image of sound overrides the evanescence that is so often ascribed to it.

This chapter opens the black box of bioacoustics by first exploring the notion that contemporary bioacoustics encourages ethical action. Because emerging sound technologies are capable of detecting small variations in sound with much greater accuracy than human listening, bioacoustics researchers can use sound to find palpable solutions to pressing social and environmental problems, rather than relying on it to understand the nature of the sonic. The scientists involved in this research must develop technical mastery of species identification, but this skill is, in fact, visually grounded since the sonic data are used as a mode for imaging in the form of spectrogram analysis. Characteristic of other sound-based research units, bioacoustics researchers are not intrigued by sound as an object so much as a method.

Since sound technologies as well as their storage devices have become digitized and automated, they are capable of capturing the sounds of global populations in real time. Sound has become an essential methodological device for identifying species and for tracking the polyphonic and polyrhythmic complexities of the various landscapes that change across ecosystems. If Henry David Thoreau celebrated the "warbling of the birds ushering in the day" (1885, p. 35) as indicative of the fluctuating quality of birdsong, current researchers similarly accept the fleeting nature of sound as "arrangements of charged particles in the semiconductive materials of solid state 'flash' memory, or the magnetic surfaces of hard drives, tapes, and minidiscs" (Gallagher, 2015a, p. 569). Geographer and sound recordist Michael Gallagher (2015b) writes elsewhere that common practices include making field recordings, including the transduction of inaudible vibrations using devices such as hydrophones and contact microphones; making compositions from field recordings, and distributing these via CDs, MP3s, vinyl, radio or online platforms such as weblogs, digital audio maps and podcasts; site-specific performances and installations; and audio walks designed for listening on portable devices whilst moving through a particular environment. (p. 469)

To elucidate the specific complexity of this proposed transacoustic community, I aim to clarify the general complexity that sound retains throughout creative imaging processes. I posit that sounding has the potential to produce interdisciplinary and theoretically innovative knowledge that enables new virtual spatializations of the earth. I proceed with a description of the historical context through which bioacoustics became a research focus for those in the biological sciences. I am especially interested in the move from recordings of individual specimens to those of whole species in their ecological contexts. I conclude with a brief return to the transacoustic community, borrowing from Jakob Johann Baron von Uexküll's (1934) notion of the *Umwelt*. Uexküll explored the making of worlds from a theoretical biological perspective; his ideas about organism self-preservation derive from a decidedly anti-mechanistic perspective, asking us to understand an organism's internal and external sense of events as its habituation to inhabiting an environment, known through existing strategies and knowledge. However, as this research is heavily dependent on the microphone as a recording device, it is worth meditating briefly on the cultural and scientific significance of the microphone in practice. It is a site of intimacy—situated, relational, and embodied.

Microphones: An Instrument of an Infrastructure

To access the affective materiality of sound-recording assemblages in relation to the scientific imaginaries of the environment, we must consider the range of practices that bioacousticians employ to animate their experiences. Not any of these instruments are simple, although the microphone is one of the more basic ones. Microphones have held a permanent place in modern mediascapes from the late nineteenth century into the contemporary context of ubiquitous computing. Whereas contemporary sound studies scholarship points to the microphone's "tympani-like" design as an imitation of the human anatomy of hearing, I prefer an approach that disentangles the mic from its predestination towards a listening event.

The microphone does not amplify: amplification requires an output such as speakers or headphones, which implies a listener over the horizon of the future. Neither is the microphone a storage unit or apparatus of capture, which would presume its attachment to a recording device. In concert with Jonathan Sterne's (2003) preference to label transductive devices as cultural artefacts, the microphone "picks up" in a manner that other technologies do not. For instance, while a camera lens may reveal patterns of light and shadow as they trace the contours of objects, a microphone (no matter what it is connected to) picks up the energy of what is alive. In this sense, the microphone does not so much live in the language of the trace (something "left over" and after the fact, a haunting) as in the vibration (which is ongoing, so has a longer temporal trajectory).

Sonic ethnographers, who track a slice of life through its soundscape, rely on such a principle to facilitate the collection of massive amounts of simultaneous movements: car engines and tyres on pavement + footsteps in layers of cadences + voices of several languages and several registers + church bells + animal sounds = a slice of the multiplicity and layers of time at a simple summer market. Just as the lens does not necessarily need to be connected to a camera for us to understand it, the microphone works according to a similar principle. Microphones raise the question of the movement behind the visual, movement behind what can be seen. Microphones engage with what is live or actual, since that which makes no movement is simply not there; thus, this is not a relationship of being or being-with, but of *being-in*.

The very point of translation from the visible to the audible helps us enter into intense, immanent, and imaginative sound worlds, a world of vibrating, churning, and almost insensible sound worlds. This point of transmission is the very spot—a virtual spot—of data collection for bioacoustics researchers. "The microphone collects all sounds indiscriminately," writes soundscape designer Hildegaard Westerkamp (2001), who claims that the microphone cannot choose which sounds it will transmit, because "it does not select or isolate them" (p. 148). The assumption underlying such a claim is that the microphone opens onto space, indiscriminately. The corollary here is that, through socialization and evolution, our ears are trained to be selective; they are, for instance, by necessity incapable of hearing the flow of blood in our body, just as those who live near airports can become desensitized to the roar of airplanes overhead. With the microphone, "Our ears are naked and open, much like those of the newborn, and can only become selective once we have begun to recognize and understand the sounds of the place" (Westerkamp, 2001, p. 148). But these assumptions need to be questioned, by virtue of the fact that there are numerous and varied microphones just as there are various cameras. Microphones, by virtue of their design, capture sounds in highly selective ways: shotgun mics, 3D mics, parabolic mics, and unidirectional mics all serve vastly different purposes.

Microphones are arranged in a variety of styles and types, but what is most important about them is their technological capacity to transduce electrical signals into sounds that are transmitted via cable (or wireless) into a sound-recording device, which itself translates those sounds back into electrical impulses. As easy as it is to say that microphones capture the sound of an environment, exactly *how* they capture the sounds is dependent upon the design and usage of the microphone in question. Some microphones must move through space to work. Some microphones sit in one spot for years at a time.

We are generally more familiar with visual forms of capture than we are with microphones and what they can catch. Indeed, enough of us struggle or have witnessed the awkward moments of keynote and conference presenters whose microphone is ignored or abused, positioned and repositioned, collapsed or dropped, only to end up to the side of the speaker's mouth—where it is indifferent to what it picks up or squeals into the air in a mighty feedback loop that causes listeners to laugh nervously as they cover their ears. Anecdotes aside, the point I wish to make is this: while microphones open up indiscriminately into space, they do not do so without the technical and social scripts necessary to operationalize them across their vast usages. Like other technologies, once they are put into proper usage, they disappear into the background.

Sound studies, science studies, and science and technology studies have not necessarily neglected the microphone. But they have not offered an exhaustive or philosophical account of its social usages outside of the professional and semi-professional music industry, or within the context of elocution in sound historiographies. Joeri Bruyninckx (2012) has written extensively on the selectivity that ornithologists have gone through in their use of a vast array of technologies to sterilize sound, including how microphones have been used. As ornithologists became more skilled at locating and identifying bird calls using technologies, and

> as listening became entangled with recording, editing, measuring, and reading, the parabolic microphone and the spectrograph represented cascading technologies of increasing control over sound, exemplified by a silenced and white background. It is through this sterile environment that the field site connects to the laboratory, as unwanted sound is intentionally eliminated. (Bruyninckx, 2012, p. 146)

A focus on sonic epistemologies elucidates certain issues of transmission, vibration, imperceptibility, and movement that visual apparatuses and visual paradigms for analysis overlook. Microphones have been central to the development of acoustic and bioacoustics monitoring. Environmental monitoring researchers, accustomed more to long-term real-time observation in a face-to-face or in situ context, are now open to the idea of longitudinal acoustic monitoring and to outsourcing of data analysis for biodiversity levels. They now organize the collection of global biodiversity data across universities, professional laboratories, citizen scientists, app developers, and wildlife acoustics. The goal is to widen the scope of monitoring a vast array of animals and creatures, which powerful microphones can assist with regardless of whether they are cheap or expensive. Further, bioacoustics researchers also use microphones in tandem with other tracking and capture technologies that are specific to their research subject. The research on nighthawks with which I began this chapter is an especially interesting area of such research, which I will return to below.

Data from environmental monitoring gathered at natural and urban sites around the globe now offer global research teams datasets that a few decades ago were unimaginable. They are connected from vast databases that have global contributors from the scientific community, as well as through citizen scientists and a spectrum of environmental activists. The microphone in a smartphone is itself powerful enough to capture the sonic emissions of a range of birds, bats, and other mammals. And microphones are now cross-referenced with location strategies, such as triangulation and GPS mapping devices. The purpose of the latter is to understand how animals populate given areas and "edge zones" such as postfire forests like the area above Fort McMurray, Alberta.

Given that the microphone plays a central role in bioacoustics and ecoacoustics research we cannot trivialize its importance. Albin J. Zak (2001), a sound-recording theorist, notes that "in many ways microphones are the technological soul of any recording project; the effectiveness of all other tools and techniques depends upon the quality of the image that the microphone is able to deliver" (p. 108). From their inception, microphones have been celebrated for picking up the inaudible vibrations that move beyond our limited capacity for hearing; such "grains" have been mobilized for our technoscientific and political advantages in anticipation of the effects of environmental change like biodiversity loss. It is what also led composer John Luther Adams to claim: "The most important musical instrument of the twentieth century may well have been the microphone" (Adams, 2010, p. 5).

As one researcher at Waterton National Parks remarked during a recording training session (Vallee, 2018), microphones "are the most important part of this package." These stale and unimpressive looking devices are fundamentally ingrained in the imagination of bioacoustics, being highly sensitive to their surroundings because of their capacity for sense and sensation. When university researchers, designers, amateur scientists, archivists, and marketing teams come together, it is difficult to say whether scientific research belongs to a laboratory, and to designate one member of these elaborate assemblages as the most important. Still, I will risk that the most important voice in this research is the voice that is recorded, the animal vocalization. I now turn to two microphone experiences that underlie the capture of animal voices, one of stillness and one of movement.

Becoming Still

Since its emergence as a field in the mid-twentieth century, bioacoustics has achieved close-up recordings of animal sounds with intimate instruments, the aim being to extract sonic information as signals for taxonomy, classification, and understanding in communication. Bruyninckx (2018) writes in *Listening in the Field* that bioacousticians have held as most prized the *recording*, deemed the epistemic centre of understanding:

> Scientists' investigations were further supported by sound archives across North America, Europe, and elsewhere, which collectively made hundreds of thousands of recordings available for bioacoustic research, analogous to the collections of physical specimens such as eggs, skins, or mounted exemplars that exist in other branches of natural history investigation. In several respects, then, sound recordings became the standard tool to document, study, and preserve the vocal behavior of birds. (Bruyninckx, Kindle Locations, Chapter 1, para. 8)

It is not so much the recording that interests me as the labour that goes into making a recording, since the recording of a nature-event requires the conscientious elimination of unwanted encounters and auditory input. Examples here include the following: human sounds such as sneezes, growling stomachs, runny noses, shifting, and tourist hikers; nonhuman sounds such as bear grunts and rustling, ground squirrels, insects, coyotes, foxes, and wolves; technological sounds such as planes and power lines; and elemental interference such as wind and rain. Thus, for close microphoning, the crystallization of the event requires a natural recording, *under the right conditions*. Microphone placement is critical in this endeavour. One sound hike in Waterton National Park required listening at 10 recording stations to record 9–11 minutes of data each time. During the recording, we were required to remain absolutely motionless—even a slight movement or a grumbling stomach interferes with the data collection, causing the collectors to start again.

The technologies of sonic capture relieve biologists and field collectors of their duty to manually identify and transcribe birdcalls, which would require of them a vast interest in song identification and the need to listen intently. Sound technologies thus liberate people from having to listen to the environment themselves, which has the effect of letting them enjoy the extended moments of dwelling in solitude, knowing (or at least believing) that the microphone is taking care of the recording and transmission of the event for later analysis. "Listening is sometimes something I get lost in when doing this," a field collector told me. "At least for the first few stations, it's very meditative. Then it becomes, well, a little repetitive" (Vallee, 2018).

This particular fieldwork encounter begins when I arrive at the research station at 2 a.m. I sleep in my car until almost 4 a.m., when the shift begins. The field collectors pull up beside me in the parking lot at about 10 minutes to 4. They are armed with canvas bags, backpacks, bear spray, bug nets, extra clothing, hiking boots, and a collection of recording equipment. This morning we will be doing live recordings at 10 listening stations. I have been permitted to join, but the team is slightly perplexed by my predominant interest in the equipment more than the animals; this focus of mine inverts the infrastructure of field recording practice.

Recording sites are indicated by orange flags. Once the recordings are made, they are placed on a computer and filed. After all of them are stored, they are sent to a contract analyst who will monitor them for bird sounds. The Waterton National Park suffered an 80% loss of the hiking trail network under a 2017 wildfire, and these sites are therefore becoming of great interest to biologists and environmental scientists alike. The supervising researcher at Waterton informs me that the shifts are long and repetitive. Each flagged station requires about 10 minutes of hiking. At the first station we wait for the sun to reveal the faces of the Rocky Mountains in a warm orange glow, followed by an emerging symphony of birds that fills the space around us as capturable data. The collectors begin each recording with a detailed description of the time and date of the recording event. We are instructed to remain absolutely still, while finding as comfortable a position as possible on mixed charred ground, dirt, rocks, and grass with mosquitoes buzzing about our ears. Remaining motionless under such conditions takes discipline.

As the Earthsong Series E3 Bioacoustic Monitor CVX Omnidirectional microphones perform their task of picking up the preferred recording event, we are forced into modes of attentive listening. The gravel and rocky ground begin to gnaw through my clothes, reminding me even more of the importance of staying still and the discipline necessary to resist the impulse to move, out of my pure creature discomfort. The underlying infrastructure of the event, the boundary of wanted and unwanted sounds, determines the movements I allow myself to make. Hence, stillness is my only option, and in such stillness my inwardness mutates into an imminent outside. In such a listening situation, my physical sense of self merges with my surroundings—I am only a trace within the larger ecology marked by the encircling signals of blackbirds, meadowlarks, orioles, and thrush, which pull my attention from internal to external. As our group attention expands, we know that this immersion is captured data. A whole matrix emerges: the acoustic community emerges as an object entangled with other objects within this ecological moment—it hangs off the air, transient yet stable, instantiating a massive, deceptively simple complexity. We have formed a rhythm.

"OK! Time. Let's move." One of the collectors breaks the silence and the moment is over. We pack up our gear, sling our backpacks on, and move to the next stop, making conversation along the way.

Recoding the Recording: Catching Nighthawks

In the context of a broader project on environmental governmentality, Jennifer Gabrys (2016) writes that animal bodies are being incorporated into a global sensor network as sensors themselves. "Organisms become computational," she writes, "both as carriers of sensors and through the ways in which their sensory ecologies are meant to provide data and information on environmental conditions" (p. 90). Animals, in this context, produce and are produced through a duality: they mediate their surroundings while they themselves are a unit of measure. That is, animal bodies open onto relations between environmental change, signalling those changes, their embodiments, encounters, and new relations that unfold. As they open onto space objectively, even while they are included themselves in the transformation of space. The animal-body-sensor is thus not static like a microphone that picks up a conversation in a room. Strapped to the animal in motion, animal sensors track spatially and temporally variable movements through space: their motility. Sensors do not pick up the environment so much as they pull into it and through it. With empirical references to projects involving the movements of badgers, seals, and white storks, Gabrys elucidates the manner in which animals *are* sensors; they are, first, technical and biological embodiments; and the technologies must evolve with the environments they sense, *as well as* the bodies they sense. This rich mesh of relations between the technical and the biological, the researcher and the researched, thus make it difficult to parse one entity from another.

Sensors, thus, open animal bodies outward and inward all at once. They open towards the changes in their environments, but inward towards the movements of the organism to which they are attached; they open the inner organism towards their environment, while placing the external into the internal. There is a challenge offered to our conception of sound here: All of these different sensors open up different spaces according to different environments: there is not one medium, not one entity, not one element, but an ongoing and porous network of bodies engaged in an ongoing life system.

Changing equipment changes the event entirely. The bioacoustics researchers with whom I worked in Northern Alberta erected mist-nets deep in the forest, in locations often only accessible by bike or all-terrain vehicle, laced with ghettoblasters emitting nighthawk calls to bait and capture them in flight. Once captured, the nighthawks are placed into small aluminium tubes and returned to the research station on the gate of their pickup truck, where they are measured and equipped with a small backpack microphone and a GPS device. The data that is subsequently recorded onto the microphone is uploaded to international research networks and measures the "sound event" of the organism (its heartbeat, its wing pace, its calls, and so forth) against the "sound-scene" of its habitat (the geophonic, anthrophonic, and biophonic data that informs the backdrop against which the sound events unfold). It was not necessarily the results of this research that interested me, as much as the infrastructural labour that went into the capture of data. This infrastructure labour, my main interest here, is emblematic of the essence of mediations that Bruno Latour (1999) describes as crossing the line between signs and things:

To be sure, we no longer portray scientists as those who abandon the realm of signs, politics, passions, and feelings in order to discover the world of cold and human things in themselves, "out there." But that does not mean we portray them as talking to humans only, because those they address in their research are not exactly humans but strange hybrids with long tails, trails, tentacles, filaments tying words to things which are, so to speak, *behind* them, accessible only through highly indirect and immensely complex mediations of different series of instruments ... Instead of abandoning the base world of rhetoric, argumentation, calculation—much like the religious hermits of the past—scientists began to speak in truth because they plunge even more deeply into the secular world of words, signs, passions, materials, and mediations, and extend themselves even further in the intimate connections with the nonhumans they have learned to bring to bear on their discussions. (pp. 96–97)

Certainly, the sonic imagination surrounding nighthawk bioacoustics (and bioacoustics generally) is not limited to strictly auditory devices, but pertains also to a larger assembly of devices intended to capture the energy and emissions of organisms in their environment. With the tools of sound capture no longer restricted to the qualities that humans have the capacity to listen for, the efficacy and pragmatic efficiency of aural-centricity has shifted. I will now briefly describe three nodes in the network of nighthawk capture I referenced above: the ghettoblaster, the mist-net, and the microphone backpack.

The ghettoblaster plays the sound of a nighthawk to bait a nighthawk. A recording's playback is neutral in the sense that it simply plays back that which it once inscribed—in this case, a stock recording of a common nighthawk's *peent!*. However, playback is also provocative in that it always happens in new arrangements, in new contexts, for new audiences, in new moments, new times, new places. The actual flying nighthawk situates the presence of the phantom nighthawk by diving towards its sound. This reinforces Michel Chion's (1999) notion of playback, in which "there is something before us whose entire effort is to attach his face and body to the voice we hear" (p. 156). Playback is about producing symmetries between subject and object, which is to say that the recording's code is "re-placed" along with the ecosystem in which it is re-placed. In other words, playback changes the place where it occurs, as the playback sounds virtually attach bodies to sounds, but in new assemblages: the nighthawk bird is a nighthawk-database. In playback, the event of recording is transformed into a new event, one that involves the bird plummeting into a "mist-net."

The mist-net, a light-weight and nearly invisible netting, emerged along with the spectrogram and was considered one of the great inventions for ornithologists. The mist-net is the silent partner of the ghettoblaster and is as important in the recording apparatus. Where once ornithologists used bait to trap individual specimens, by 1947 they were placing mist-nets around the periphery of their observation areas. Much like the microphones that capture everything that they come into contact with, the mist-nets captured everything that would attempt to pass through them, which allowed a more realistic impression of the numbers of specimens occupying a zone. Significantly, they led to the rise in quantitative measurements of species over qualitative descriptions of them.

While today many bioacoustics researchers consider the ARU the best practice for obtaining sonic data, they still use the mist-net to capture and track individual specimens (as part of a group), and to receive a more high-definition image of a population as it occupies a territory. However, parsing the differences between qualitative and quantitative research is not meaningful to biodiversity assessments. Instead, the entire process counts as the labour of a methods approach, which is currently gaining popularity in interdisciplinary research. Michael Mair and colleagues (2015) write of the fabled divisions that keep quantitative and qualitative research methods divided, suggesting that such a practice is best avoided because

> labelling research practices as qualitative or quantitative (or indeed "mixed") may well have some uses (as badges of membership, for instance), but the labels themselves are not specifically descriptive of those practices and should not be treated as such. Knowing whether a piece of research is qualitative or quantitative, interpretive or calculative…is much less important for characterizing that research than understanding the specific ways in which it makes "the social structures of everyday activities observable"—that is, how it puts society on display. (p. 54)

The researchers, upon capturing the nighthawk, stuff them into tubes to bring back to the truck, where they record their standard physiological measurements. The final stage in a well-choreographed series of measurements is to strap a backpack microphone between the bird's wings: this is a specific way to record the nighthawk's vocalization, as it moves through particular habitats and ecologies. A relatively recent invention, the microphone backpack gives equal weight to the voice of the bird, its bodily vibrations and movements, and the spatial distributions of sounds in its surroundings. This sound-scene analysis (Stowell et al., 2017) is intended to create a portrait of an individual animal, something far more difficult to do with static microphones. The microphone backpack, in contrast, gathers an individuated image from the individual organism's unfolding environment. In the past it was not possible to get such an individual picture.

The voice stands as a kind of extension towards the general physiology, the wing patterns, vocalizations, heartbeat, breath, external movements, elements, and environments. There is no one voice that is not many. Additionally, the microphone backpack tracks animals in otherwise difficult-to-access situations allowing for intimate analysis in a range of environments that present physical and sonic obstacles. Thus, the sound-scene analysis (also called *sound-event detection* [see Mesaros, Heittola, & Virtanen, 2016]) is the primary mode for analyzing the collective movements and sounds, and for dividing them onto a spectrogram. One researcher played a recording on their laptop: the wind made a sound like the scratches on a nineteenth-century phonograph cylinder, with an undulating 3–5 beat pattern, which I learned were the wings in flight, and an occasional *peent!* that pricked through the drone of the wind. The researcher started to smile at the sound, as though they had just listened to a long-lost recording from their past.

All these sonic characteristics belong as objects to the multiplicity of the bird; not one alone cannot represent the bird, for the bird is in concert with its surroundings. We are accustomed to bringing an organism's organ—such as a voice or a larynx—to light through isolation, instead of in tandem with its ecology. But every insight about a voice has emerged within a context of multiplicity. Researchers on the voice and the laryngoscope knew that voices were an effect of vibrations in the throat, even before those vibrations found their destiny in the actual mirror of the laryngoscope. Those vibrations were not yet proven to exist, but were projected to reside in an unknown space of the voice. In this sense, once it was discovered, vibration became the rubric through which the voice was set free of the body in the process of recording—for the needle of the stylus worked according to the principle of vibration that was also in the science of the voice. Vibration thus became the energy between the voice's capture in the human body and preservation on the recording assemblage. So the organism is entangled in a soniculture assemblage, never independent.

These nighthawk recordings are non-localizable recordings given that they reflect the sonic emissions of birds engaged in global migration patterns, which by definition means the birds pass through numerous locales. Further, they evidence that this species faces serious population declines. This relationship between the sounds and the spatial contexts within which they occur calls up the spatiotemporal network to which sound belongs. This network goes beyond the physical body into a massive virtual body of multiplicity that links many nodes: the technology, the organism, and the researcher and their joy in the pure sound of the nighthawk, as well as ideas and knowledge about sound and about the species.

Thus, despite the current proclivity to treat sound as a scientific object divorced from the sense experience we have of it, I am more interested in it as a soniculture assemblage, a part of a nature/culture/sonic assemblage. Science may determine nature as an object, but even when scientists take this tack, they cannot diminish the multiplicity of any assemblage, which arises from its entanglements: the nighthawk/mist-net/backback assemblage is entwined with the nighthawk-ecology-assemblage and the researcher-listening-smile assemblage, in ways both embodied and spatialized. Yes, it is at once captured, made sense of, but through this sensemaking, multiplicities converge and become incarnate in unexpected places.

After one of the researchers with whom I worked strapped a backpack microphone between the wings of a captured nighthawk, she released it to discover that it dropped like a stone. The microphone was not properly installed and was causing a disequilibrium in the nighthawk's capacity for flight. Swiftly she moved to its writhing body on the ground to remove the device, which she did with ease, and backed off to watch the bird fly into the inky darkness. As the microphones take a great deal of labour and care to instal, it comes as a disappointment when they do not work. In this ethnographic fieldwork, sound is connected to vibration, but more importantly, sound is method, with conservation the goal. This raises fresh questions about the philosophy of soundworlds and the worlds between sounds, which requires a turn to the transacoustic community as a way of imagining the imaginary organ of place.

The Transacoustic Community

My proposed concept of transacoustic community differs from the concept of the acoustic community. The acoustic community is complex and information-based "in terms of sound, and therefore sound plays a significant role in defining the community spatially, temporally in terms of daily and seasonal cycles, as well as socially and culturally in terms of shared activities, rituals, and dominant institutions" (Truax, 2001, p. 66). Transacoustic communities, on the other hand, respond to the necessity for new trans-species community formations, bound by both the rigours of scientific research and the ethics of love and compassion on a local and global scale. Transacoustic communities appreciate the entangled nature of researchers, their instruments, biotic forms, and their significance for human and nonhuman environmental issues. This concept elucidates just how the biotic/acoustic/technic/plastic assemblage of nighthawk + speaker + mist-net + backpack activates a system of *relatings*.

Unlike the acoustic community, which is brought to us through the neutral deployment of microphones and the broader recording apparatus, the infrastructure produces a relational ethic that is beyond data-driven science, producing affective attachments such as the naming of the birds. Though the research is young, researchers are trying to understand how individual animals emit individual vocal cries beyond the general vocalizations of their species. To fuse the hypothesis with human-bound sentiments, they name the birds: Tom Waits, David Bowie, Gord Downie (who, I was told, was found under a rainbow at dusk). Nighthawk research, then, activates the boundaries of plastic, technics, and biotics as they are entangled in one another in the research process, but these boundaries are maintained through massively complex relationships.

The common nighthawk, the bioacoustics researchers, and the technologies through which they are measured and made sense of globally and locally exemplify the image of a transacoustic community, which is bound by an elaborate recording/playback apparatus that is not necessarily reducible to the listenable, but instead expands more generally into recording as a technical and cultural set of images. Entities open through their surfaces and edges onto other entities. These openings, to which variegated routes of access exist, are precisely the point of interrogation for bioacoustics researchers. I have set out in this chapter to explore some of these access routes. Of course, there are many ways of doing bioacoustics research, but they all converge on the creation, maintenance, and breaking through of an entity's contained space, through its sonic emissions. Technological assemblages belonging to bioacoustics researchers are intended to create new images and imaginations for how these breaches are achieved.

Entities sound. And in that sound, they make up their worlds. But since entities sound out to other entities, the worlds to which these first entities belong are certainly not contained. This notion that a world is not self-contained, but rather porous and protean, makes it necessary to interrogate the underlying function of worlds as open. Uexküll's philosophy of the *Umwelt* that I referenced earlier translates literally as "world around" (Brentari, 2015, p. 75), where an organism's boundary is shaped by habituation to the environment. Elizabeth Grosz (2008) has expanded on this position, explaining that in the human world the architect is the equivalent figure—this professional brings together things at the demarcation of boundaries, creating heterogeneous expressions within a space that are given meaning through their very heterogeneity.

Grosz accounts for the famous tick that appears early in Uexküll's book (1934/2010), *A Foray into the Worlds of Animals and Humans*. Uexküll argues that animals are embedded in affective worlds. Ticks, for instance, use the smell of chemicals, the heat of the sun, and the flesh of the mammal to complete their worlds, achieved once they have conjoined with another organism, such as the mammal; that is, their world is defined by their connection to another's world. The tick's world is thus complete when it attaches to the edge of another world (the world of the unaware mammal, for instance, whose own totality the tick is equally unaware of). While the organism's *perceptive world* is an inherited, species-specific conscious perception of those objects that it perceives as outside of itself (such as the sound event mentioned above, the *peent!* of the nighthawk), the organism's *operative world* completes its immersion in an environment by merging with it. For example, as in the sound-scene, the flutter of moth wings draws the nighthawk's dive into it so that the bird can consume this insect food (see Brentari 2015, p. 99).

At stake here is thus the indeterminate nature of worlds as they forever open onto and into other worlds, along with the maintenance of worlds as highly dependent on the membrane of their milieus. Grosz (2008) writes that such worlds are *musical* (a common audio-based resource for those constructing idealized collective experiences):

> The music of nature is not composed by living organisms, a kind of anthropomorphic projection onto animals of a uniquely human form of creativity; rather, it is the *Umwelten*, highly specifically divided up milieu fragments that play the organism. The organism is equipped by its organs to play precisely the tune its milieu has composed for it, like an instrument playing in a larger orchestra. Each living thing, including the human, is a melodic line of development, a movement of counterpoint, in a symphony composed of larger and more complex movements provided by its objects, the qualities that its world illuminates or sounds off for it. Both the organism and its *Umwelt* taken together are the units of survival. Each organism is a musician completely taken over by its tune, an instrument, ironically, only of a larger performance in which it is only one role, one voice or melody. (p. 43)

At what point does a boundary turn into a breaking point? Or, at what point does the edge of one boundary merge into the edge of another? Uexküll establishes his position against physiological accounts that would see organism and inter-organism behaviours as effects of stimuli reactions between different parts of an organism. Such accounts position organisms as isolationist, against their environment; further, they resist the notion that an organism possesses agency in the construction of its world and needs agency to converge with the edge of another world. Therefore, the organism does not create its world but, rather, its co-constructive capacity for going into other worlds.

Every melodic contour is an edge of another's world. It is sound which accounts for the breakthrough between the edges of worlds. Sound, as I present it here, is the multiple transduction of individual energies from separate worlds, involving at once the assemblages of vocal tissue and environmental biotic and abiotic movements, including all other bio-phonic, geophonic, and anthrophonic sounds. When software has been programmed to detect a variant in sound, a transduction, this sonic infor-mation is registered significantly more accurately than with organic lis-tening and pattern identification (though researchers often test-listen to samples to assure general accuracy). The use of this software introduces technical or mechanical listening (that is, inorganic) into the process of exploration and discovery. Longitudinal studies are often multisited and multimicrophoned, and record more sound (which is unstable and in flux) than it is possible to listen to organically, thus generating highly individ-ual outcomes. Most important, they crystallize a node in a transacoustic community.

Imagination, to imagine, to image: these variations on a term point to the slippage (linguistically and otherwise) of image. Countless philo-sophical explorations of image and imagination can be found, but the question of how images come to be made through sound is quite another matter—one that is often grounded in routine empirics and the tech-nics of observation. To reiterate the essence of transacoustic communities: bioacoustics researchers are less interested in sound as an object of analysis itself than they are in using sound (and expanding definitions of sound) for high-definition insights into environmental and social problems. As such, they open up sound to a cross-modality of senses and transductions.

We need to further think through sound as an emerging complexity with expanding boundaries. For instance, the world comprising the ghet-toblaster and the mist-net releases us from thinking of a recording as a pristine reproduction of sound "as it is." Instead, we think here through the image, imaging, and imagination of sound, as it affords the heard and the unheard; within and beyond, below and above normal human hearing, the capacity to live in resonance between objects and entities like ghetto-blasters, mist-nets, nighthawks, and the global research teams that coalesce. There is more at stake than having the potential to capture more sound at a higher definition with a wider grasp at a longer rate, as the opening exam-ple of the common nighthawk suggests. To the extent that we imagine place as a boundless mediation through and resonance between techno-logical interfaces—we have a perspective that privileges the aesthetic and technological intervention over the usefulness or pragmatics of learning about biodiversity preservation. We could instead consider in what ways new digital technologies reveal that the image of sound is not mandated to be "listenable." The capacity of massive data storage and the live-response possibilities of broad transacoustic communities may, for example, lead to sound imagery that presents the transduction of sound energy into forms of information that, above all, meet the needs of research.

In closing this chapter, I note that some critiques exist against the tendency to entangle animals in cultural theory, thus claiming success for decentring the exceptionalism of the human species. For example, Matthew C. Watson (2016) makes a timely rebuttal to the root metaphoric assumptions that are unquestioned within naturecultures: namely, that poststructural terminology for animal behaviour place an unchecked "mytheme" around the animal, in the service of opening up science. This removes species from their interconnected phylogenesis and their adaptability. Important scientific advancements in extinction intervention are diminished in place of metaphors that paint the impending disaster with a "cruel optimism" (Berlant, 2011, quoted in Watson, 2016). Watson notes, "Those of us in the human sciences must be willing to take seriously the statements that biological scientists have about their objects of knowledge" (p. 166). I argue instead, based on the empirics of bioacoustics research and researchers, that sound is an extractive measurement for the purposes of data analytics. But within these data analytic processes, there remains the embodied, the aesthetic, the affective, and the political. Emerging technologies therefore contribute to the decentring of anthropocentric hearing, but they recentre the relations that humans may have with other species, as we will see in the next chapter. We will now turn to the becoming computational of sound.

References

Adams, J. L. (2010). *The place where you go to listen: In search of an ecology of music.* Middletown, CT: Wesleyan University Press.
Berlant, L. (2011). *Cruel optimism.* Durham, NC: Duke University Press.

Brentari, C. (2015). *Jakob von Uexküll: The discovery of the umwelt between biosemiotics and theoretical biology*. New York, NY: Springer.

Bruyninckx, J. (2012). Sound sterile: Making scientific field recordings in ornithology. In T. Pinch & K. Bijsterveld (Eds.), *The Oxford handbook of sound studies* (pp. 127–150). Oxford, UK: Oxford University Press.

Bruyninckx, J. (2018). *Listening in the field: Recording and the science of birdsong*. Cambridge: MIT Press.

Chion, M. (1999). *The voice in cinema* (C. Gorbman, Trans.). New York, NY: Columbia University Press.

Gabrys, J. (2016). *Program earth: Environmental sensing technology and the making of a computational planet*. Minneapolis: University of Minnesota Press.

Gallagher, M. (2015a). Field recording and the sounding of spaces. *Environment and Planning D: Society and Space, 33*(3), 560–576.

Gallagher, M. (2015b). Sounding ruins: Reflections on the production of an "audio drift." *Cultural Geographies, 22*(3), 467–485.

Gill, L. F., D'Amelio, P. B., Adreani, N. M., Sagunsky, H., Gahr, M. C., & ter Maat, A. (2016). A minimum-impact, flexible tool to study vocal communication of small animals with precise individual-level resolution. *Methods in Ecology and Evolution, 7*(11), 1349–1358.

Grosz, E. (2008). *Chaos, territory, art: Deleuze and the framing of the earth*. New York: Columbia University Press.

Hutto, R. L., & Stutzman, R. J. (2009). Humans versus autonomous recording units: A comparison of point-count results. *Journal of Field Ornithology, 80*(4), 387–398.

Laiolo, P. (2010). The emerging significance of bioacoustics in animal species conservation. *Biological Conservation, 143*(7), 1635–1645.

Latour, B. (1999). *Pandora's hope: Essays on the reality of science studies*. Cambridge, MA: Harvard University Press.

Mair, M., Greiffenhagen, C., & Sharrock, W. W. (2015). Statistical practice: Putting society on display. *Theory, Culture & Society, 33*(3), 51–77.

Mesaros, A., Heittola, T., & Virtanen, T. (2016, August). TUT database for acoustic scene classification and sound event detection. In *2016 24th European Signal Processing Conference (EUSIPCO)* (pp. 1128–1132). IEEE.

Sterne, J. (2003). *The audible past: Cultural origins of sound reproduction*. Durham, NC: Duke University Press.

Stowell, D., Benetos, E., & Gill, L. F. (2017). On-bird sound recordings: Automatic acoustic recognition of activities and contexts. *IEEE/ACM Transactions on Audio, Speech, and Language Processing, 25*(6), 1193–1206.

Sueur, J., & Farina, A. (2015). Ecoacoustics: The ecological investigation and interpretation of environmental sound. *Biosemiotics, 8*(3), 493–502.

Thoreau, D. (1885). *The writings of Henry David Thoreau* (Vol. 6). Boston: Houghton Mifflin.

Truax, B. (2001). *Acoustic communication* (Vol. 1). Westport: Greenwood Publishing Group.

Vallee, M. (2018, September 17). *Tour de Frank.* Field notes at the Frank Slide site.

Viel, J. M. (2014). *Habitat preferences of the common nighthawk (Chordeiles minor) in cities and villages in southeastern Wisconsin* (Doctoral dissertation). Retrieved from https://dc.uwm.edu/etd/516/. The University of Wisconsin-Milwaukee.

von Uexküll, J. J. (1934/2010). *A foray into the worlds of animals and humans: With a theory of meaning* (J. D. O'Neil, Trans.). Minneapolis: University of Minnesota Press.

Watson, M. C. (2016). On multispecies mythology: A critique of animal anthropology. *Theory, Culture & Society, 33*(5), 159–172.

Westerkamp, H. (2001). Speaking from inside the soundscape. In D. Rothenberg & M. Ulvaeus (Eds.), *The book of music and nature: An anthology of sounds, words, thoughts.* Middletown, CT: Wesleyan University Press.

Zak, A. (2001). *The poetics of rock: Cutting tracks, making records.* Berkeley: University of California Press.

5

Sounding Data

Elemental Sounding

A long listening hike can usually take me out of a writer's block. At the time of writing this book, I lived in a remote area of Canada in Southern Alberta—the Crowsnest Pass—which is walking distance from creeks, rivers, and mountains, surrounded by wildlife and the vast space of a mountain range. Noticeable on my long walks were the birds; something like 160 species of birds live in this area, which also holds a rich depository of fossils that are routinely exposed due to the unstable ground upon which the area is situated, namely, the edge of the Continental Divide. Throughout the day, red-winged blackbirds, nightjars, ravens, crows, and yellow warblers fire their calls and their songs through the trees, more audible than visible.

One hike I particularly like winds up a small but steep incline with a view of Turtle Mountain. The north face of this mountain was defaced by a massive rockslide at the turn of the twentieth century that killed the mining residents of Frank, Alberta who had built their community at its base (see Chapter 6). The sight of the Slide commands stillness, an ominous silence that is narrated by the emerging symphony of birdsong. From this particular viewpoint, you can see the mountain and its rubble

© The Author(s) 2020
M. Vallee, *Sounding Bodies Sounding Worlds*, Palgrave Studies in Sound,
https://doi.org/10.1007/978-981-32-9327-4_5

of destruction below. It is from this view too that the sound emerges from within the habitat: the feverish caw of crows, the agile *chicka-dee-dee-dee* of the chickadee, the snarky *witchity-witchity* of the yellow warbler—all above the soft hush of the Crowsnest Highway (Highway 3) that cuts through the remnants of the rockslide below. The highway is dangerous, routinely the deathbed of many deer, cougars, moose, and bear who have lived in and rely on the territory as a natural wildlife corridor. Drivers sometimes equip their vehicles with wind-driven whistles intended to warn the wildlife of any oncoming danger, including themselves. One can occasionally hear these whistles, uselessly whining like an alarm bell taped to a nuclear missile.

Hiking in this area, these sounds of the outside are subsumed by my own deep breaths, brought about by the steep incline to my favourite listening point. I slow my breath so as to absorb more of the information outside. In slowing down, the sounds around me emerge more fully, introducing more songs, occasional animal steps, trees that creek in the wind, the faint hush of one of the many nearby rivers and creeks—and I can begin to place them. When the wind gets going the trees rub together and make an intimidating sound, which can feel consuming.

And, so, in listening, I disappear. I listen with my whole body, my eyes directed by sound towards openings in my visual field. I never close my eyes, never miss the chance to see a bohemian waxwing, a blue jay, or a common wren. In this intense listening, my body vanishes as it becomes a silent mediator in the process of opening onto the information from the outside. It is difficult to even say whether this is my body, or whether, caught up in the ecology of listening, my body is caught up in the surroundings that contain me.

Sounding is the orientation towards sound exemplified in these moments. This sort of approach to sound is that *very preparedness for listening* that precedes hearing—thus, this sense of sounding captures the notion of *emergence*. Listening, as it is articulated by Les Back in *The Art of Listening* (2007), is a central practice for scholars across the humanities, the social sciences, the natural sciences, and the arts alike. However, listening carefully means considering an entire context, "a listening for the background and the half muted" (Back, 2007, p. 8). Listening in an ecological context thus takes into account a contemplative, immersive, and aesthetic approach to the orchestra of nature. But the openness that precedes listening is what *sounding* is all about—it is everything that goes into the event of listening. Sounding is to listening as emergence is to event.

Listening has an obvious embodied dimension, and this kind of embodied listening goes only so far before I want to record the sounds to capture the atmosphere. As I alluded to at the start of this chapter, I have a long-standing interest in recording things—not just music, but events, gatherings, and festivals. But listening also contains a technical, highly mediated, and digital dimension, one that requires a *reaching out* or an opening of this immediately experienced, embodied, and multidimensional world. This extension links to dimensions of recording and transduction, and also to sharing, community, and interspecies collaborations, some of which I have touched on in previous chapters.

What follows is an intentionally broad discussion of bioacoustics as an example of how emerging technologies used for scientific research can act as an empirical event against which theorizing is possible. This is less a "theory of science" (or field of science) than a call to ask questions through the lens of the sciences that vitalize new theorizations. Bioacoustics, thus, should be expedited not only into the scientific body (which it has) but into the language of cultural theory as well, since its tools and technologies expand the borders, boundaries, and flows of knowledge and transdisciplinary research innovations. The chapter explores several aspects: the necessity and usages of bioacoustics research; its co-emergence with technologies of sonic storage and dissemination; its correlative discoveries; the implicit retangling of hearing and listening; the places where animals are recorded; data compression; expanding edges and boundaries of conceptions of ecosystems; and finally, the new soundscape ecology. The scientific perspective helps us understand how we hear just a slice of the sonic spectrum, and that sound is a method for elucidating how researchers are implicated in their research, including the animal/human/sound/species/techno/listening semiospheres that contain and expand their capacities and potentials.

Embodied Infrastructures

Intent on capturing some of the mobile serenades I hear during my hikes, I had downloaded a relatively inexpensive app for my iPhone, Song Sleuth, which records and automatically identifies birdcalls. This app introduced automated animal detection software to the market for both professional and amateur research usage. With Song Sleuth, I can record birds, identify them, and send those recorded sounds (including a GPS coordinate) to others via email or messaging, connecting my embodied and highly personal experiences to a global network of bioacoustics researchers and amateur bird listeners like myself. The app uses simple bioacoustics technology, an efficient detection tool for gathering early warnings about species in need of conservationist intervention. Tracking the birds makes it easier for global researchers to develop a "big picture" of populations at risk, migratory patterns, and mate selection behaviour. Thus, the simple act of listening—simply being still and paying attention to the surrounding sounds—can have an immediate and long-lasting impact on professional scientific research.

Academic and professional research teams in bioacoustics consist of many funded researchers, who collect and analyze data, and disseminate their research findings. However, citizen scientists, like myself, who upload data recorded during leisure hours are now seen as key players in widening research teams on a global scale. Research teams that make considerable use of data from citizen science, such as Cornell University's Macaulay Library, corroborate that research data from public contributions arrive at a much faster rate than ever before.

Regardless of their scientific contributions, citizen scientists also experience a great deal of health benefits: they enjoy the privilege of sensing the life of animals and insects whose survival is under threat and whose presence is not readily apparent in everyday life. Children, especially, experience enriched connections with nature, and adults spend more of their leisure hours moving. Citizen science thus contributes to a non-sedentary lifestyle.

Aside from good health, some researchers praise citizen science for contributing to a growing sense of awareness about environmental problems. However, some researchers argue that awareness is, while certainly an ideal outcome to citizen science, a paradoxically challenging one to measure. Various studies have, however, proven that sound-based methods of empirical exploration, such as bioacoustics, are linked with a meaningful awareness connected to space. Thus, including citizen scientists in bioacoustics research is a practical and cost-efficient means of including global and local populations in a method that contributes to spatial (and, by extension, environmental) awareness.

Sonic technologies, as part of this citizen science, are opening up new modes of citizenship and belonging, new relations between humans, non-humans, and the technologies that mediate their information gathering and knowledge building. A principal axiom of citizen science is that of participation, where furthering knowledge comes as its own reward, and where the other side of participation (passive observation) is stood on the defensive. Citizen science rewards its scientists with the great outdoors, solitude, and the vicissitudes of nature (Lawrence, 2006, p. 292), while proffering a set of habits that can bind together the sometimes disparate elements of everyday life, knowledge, and nature. But citizen science is also an experience that can grant cultural capital (Bourdieu, 1984), as it is a field comprised of those who consume the narratives of the Anthropocene, and who, through these narratives, produce their own exclusions. By acting ethically, citizen scientists are excluded from the "rest" of the human race—the trip from bystander to participation. Citizen scientists are privileged through their many responsibilities: to their community, to research teams, to government, as well as to social media. Anna Lawrence's (2006) figure of dynamic interactions in Voluntary Biological Monitoring depicts the interrelations between action, experience, data, and policy that reflect this dynamic. The citizen scientist is thus presented with a range of responsibilities that are embodied and cognitive: to understand the world as an interconnected mesh (Morton, 2010); to understand that in every passing present looms an imitable and potentially catastrophic future; to become creative in their means of data collection and data sharing; to participate in a political awakening about the rhizome of human and non-human agents (Latour, 2005); and to commit to immersing themselves in a field.

But it's also often assumed that "citizen science" rests on a deficit model in the sense that people are active in part to achieve recognition for contributing to knowledge and building community. Such a perspective reflects the earliest developments in sustainable ethics (Irwin, 1995): that a person's inner worth is validated by their "new relation" with a sustainable world. This is an old linkage between self-worth and unsustainable practices that are treated as though they are pathologies to be corrected through expert knowledge—but this would be a radical biopolitical reading of citizen science. Citizen scientists are encouraged to use technology to better their own and the larger scientific knowledge, to develop broader, nationally expanding data collection archives, and contribute towards a better global future. Given that technologies are *relatively affordable* to the general public, the benefits of citizen science are brought within the orbit of science and knowledge mobilization, as well as multiple and immersive modes of data collection.

Certainly, the citizen science culture and its inclusionary economy place pressure on everyday people to live according to their duty to nature. Those who dispose of their recycling, drive excessively, shop at WalMart, or let balloons go into the sky as part of a memorial service (all activities I have recently taken part in), are not simply ignorant, but are earth-citizens in utero. In particular, "participatory citizens" engage in all sorts of earthbound activities:

- Taking photographs while sea-diving (Latimer, 2013).
- Sharing their data through social networks (Lievrouw, 2010).
- Forming ties with the scientific community in traditional peer-review publications.
- Being credited for contributory, collaborative, and co-created knowledge about noise pollution, monarch butterflies, volcanoes, soil samples, marine phytoplankton, redwood forest plants (Connors, Patrick, Lei, & Kelly, 2012).

Customized citizen science projects that such individuals participate in happen in diverse locations and contexts: the zoo; home; at night; in school or sports stadiums; at the beach; as emergency responses; on the internet; in oceans, streams, rivers, and lakes; in snow or rain; in the car; on a hike, walk, or run; or while fishing (see scistarter.com).

However, some argue that citizen scientists are more accurately described as co-producers of conservation habits, rather than knowledge contributers (Cornwell & Campbell, 2012). Nonetheless, data gathered by such individuals are being incorporated into the traditional university system of research, which move the research process to research dissemination more transparently. Overall, the multifarious ways that individuals may engage in citizen science have the effect of shifting their subjectivity away from the reflexive and towards the worldly.

But these are more than citizen science initiatives, since they too are entangled in new digital mediations, interfaces, and technologies of sense; Jennifer Gabrys (2016) has called these new initiatives "citizen-sensing" projects. By displacing the citizen scientist with citizen-sensing, the citizen can now connect directly to research initiatives with simple hand-held devices, thus being directly implicated in the interventions, corrections, and policies that scientific research brings to struggling ecosystems and depleting animal populations. According to Gabrys (2016), citizen-sensing incorporates into the hands of the everyday user those sensing devices that have become otherwise ubiquitous in contemporary urban and rural monitoring. Notably, the implication of this integration into subjective experience is that the citizen scientist approaches the environment and ecological habitats in an increasingly computational manner—in that they do not need to rely on their own senses, but rather the senses of their devices—to accrue data and to upload it to monitoring projects online. The result of this shift in attention, which itself derives from an encroachment of ubiquitous and peripheral media into the experience of the user, is that users experience a new oscillation between new proximities to their environment, but new alienations from their environment as well. While they are not engaged through their senses, their increased mobility and the ease and efficiency of the projects supports more fruitful research endeavours.

Gabrys (2016) argues that while the question of direct engagement lies at once with the technologies, those digital technologies open onto objectively discoverable new worlds, while themselves constituting a new world by opening to a new relation between citizen scientist, citizen-sensing, and nonhuman bodies. Thus, the nonhuman bodies mediate the information. A number of new theoretical challenges to media and cultural theory exist as a result of such a shift, Gabrys suggests: increasingly thorough readings of environments, new global mappings, animal mobilities, community organization, increased awareness, direct policy, and implications for action.

Developments in citizen-sensing have made for a more proximate relationship between people and nature, but in an unexpected way: instead of an appreciation for nature from the perspective of an admiring subject, people are more closely connected to the global research importance of ecologies. They can grasp the interconnectedness of habitats across the globe, which are understood through grounded scientific research. My crow does not live in the Crowsnest Pass. He lives all over the world, a taxidermied code alongside others in a global show. With the connections between human and nonhuman actualized through the digital network, the capacity for caring about those connections may increase as well. Gabrys writes: "Sensor-generated ecological data is often gathered with the purpose of articulating more exactly the scale and details of environmental change, but here monitoring extends to include other types of citizen interventions. These watchful humans are seen as vigilant 'live eyeballs,' as well as caring participants" (2016, p. 79). Indeed, while computational media closes aesthetic contemplations of nature by supplementing listening, it opens up new ethical behaviours of intimate caring.

An inseparable component of embodied listening, especially in sound and soundscape studies, is its imbrication with recording apparatuses. In the intersection between embodied listening, the ethics of attuning oneself to an ecological atmosphere, and recording technologies, the act of recording generates a delay between hearing and listening. Recording, then, is essentially a space between hearing and listening, which suspends the latter. That is, as Barry Truax (2001) reminds us, while *hearing* is scientific, referring to the range and frequency of sound one is capable of detecting, *listening* is interpretive, involving a complex connection between the acoustic information and the social and cultural milieux. Recording opens listening onto a multidimensional, potentially infinite, array of interpretive situations.

In resonance with authors who hold that embodied listening is deeply entangled in the sociotechnical imaginations of history and social structure, sound recording therefore brings a strong ethical and historical dimension to the capturing of the event, as it was described in the previous chapter. However, unlike recording devices that record onto a medium to archive and play back at another date for a human listener, digital recording devices such as Song Sleuth perform a different function. They are uploaded and tracked on a giant database of sounds, and tagged with GPS technology that allows for the visualization and global positioning of the sound. They are readily available for other users to listen to and download for comparison, and to determine what animal sounds are where. Recording now holds a new authority over the previously privileged status of human ears that were conditioned to contemplate, wonder at, and hold in awe the sounds of nature and of the earth. Indeed, our relationship with all the actors in a sounding event are informed by the way we store sound, archive it, and put it to use later.

That a high quality recording of a crow can be made using a smart phone, and that such a recording can be uploaded immediately to a central database, represents a massive shift in how sound-based research is conducted, who might be involved, and what sorts of communities may grow from and around these technological shifts. One thing that has changed drastically in a digital recording is the deep time that it generates; as was seen in Chapter 3, sound recording can take us into a pure past, a shared ancestry on the bud of Darwinian evolution. Chapter 4 explored how sound can take us into a virtual future, with an impossible amount of sonic information that is intended to model which habitats are in need of conservationist intervention. Consistent with understanding the qualitative changes that have come about in changes to sound technology, this chapter intimately connects the embodied present with the pure past and future potential.

Animal Sounding

There is a longstanding interest in transcribing animals' sounds: Anthanius Kircher's *Musurgia Universalis* (1650/1970) imagines connections between humanity and nature by assigning certain animal songs to the cosmos, such as the sounds of various birds as well as the sloth, which exhibits a six-interval vocal range. Ludwig van Beethoven, through his well-known wilderness walks, incorporated birdsong into his compositions, with the most obvious instance being his Symphony No. 6 that includes "cuckoo calls" (see Baptista & Keister, 2005). Modernist French composer Olivier Messiaen (1908–1922) transcribed birdsong and manipulated their speed and range to capture an otherwise imperceptible dynamic (Hill, 2007). Contemporary musical improvisor David Rothenberg's (2008) well-known live and recorded performances for and with a variety of species continues this tradition of linking aesthetic, sound, sense, imagination, and collaboration. Most of these, and other examples, have relied on an aesthetic of listening that nineteenth-century music critic Paul Scudo once referred to as "the divine language of sentiment and imagination" (Scudo, cited in Johnson, 1995, p. 272).

I am interested here in the imagination that belongs to the biological sciences. Before the mid-twentieth century, bioscience researchers centralized listening in their data collection and analyses, transcribing the sounds of animals for the purposes of discovering keys to biodiversity, mating behaviour, and the anticipation of biological change. Notably, because recording devices were too cumbersome, researchers had to rely on having a musical ear to transcribe sound in an onomatopoeic form as I have done elsewhere with the *peent!* of the nighthawk. Unconvinced by this method, Albert R. Brand (1937), at Cornell University's Ornithology Research Lab, had attempted to capture bird song with "sound film" (used otherwise for Hollywood "talkies"), which captured both the image and the sonic emissions of birds. He considered this a more objective means of saving sound, as he writes:

> [R]arely do two observers hear the same song in exactly the same way. The song is not noticeably different when produced by varying members of the species, but by the time the sound waves have affected the listeners' hearing apparatus, and have been transferred by the nerves to the brain, and interpreted by that organ, it has created an entirely different sensation and impression on each individual listener. (Brand, 1937, p. 14).

Although Brand's films were grounded in visual images and movement, he was still keen on viewers listening in real time to the sounds of animals. However, by the mid-twentieth century, the spectrogram was introduced to ornithologists to visualize sonic information. Scientists no longed needed a musical ear, but rather, technical knowhow. Spectrograms helped to democratize access to sonic data, sound analysis, data collection, and made contributions to science throughout the late twentieth century. Today, the spectrogram image (and its variations) is the most common image of an animal's utterance. This image is inseparable from a new kind of work, which frees the scientist of the burden of listening; instead, their attention is shifted first to placing the equipment and, then, to using it to capture the animals' sounds. The researcher, now liberated from their own ear, works with technology that can pick up the transmission of information. Better yet, the spectrogram is equipped with a capacity for accurate visualization, given that the vibrations from the needle on the machine are etched into a paper surface. Significantly, the spectrogram caught more than the sound of the organism—indeed, it collected data on the whole situation within which it was situated. This transcription of the atmosphere, of the organism's world, allowed researchers to visualize the polyrhythmic complexities of its environment, including its communications with other species.

The spectrogram demanded a unique, visually grounded art of its own: calligraphy, traced on paper, meticulously teased out the upper portion of the recorded sound so as to discover an arc represented through space (frequency) and time (duration). Birdsong would no longer be described using words, or onomatopoeia for that matter, but had to have a direct *inscription on to pages, using ink and paper, of the ecology in which the organism was situated.* These various mediations and transductions were less contiguities, points of contact, than they were direct feelings, motivations, and movements onto the page to trace the otherwise invisible (but real) contours of the bodies responsible for producing them. This calligraphy demanded particular, visually detailed sonic information that reduced the need for the humans to be involved in its production, though in the past this information may have been achieved by someone listening attentively and tracing the contours of a sonic inscription to identify an organism or its behaviour.

The spectrogram was adept at picking up certain important information—the environment and the ecology in which the animal was situated—whereas the onomatopoeic transcriptions isolated song from context. What became more important, then, was less the taxonomy of the animal than what the animals' sounds could tell researchers about their surroundings and their environments: that is, how they were situated within a community or a sound ecology.

Instruments retangle relations between animals, their sounds, and researchers through emerging sound technologies, and account for new motivations for research, new questions, new methods of sonic analysis, and new, increasingly imperceptible ways of listening to animals' sounds. These relations collectively justify the need for developing a framework for understanding the actions and interactions of organisms, human and otherwise. And such connections can be thought of as "agential," in the sense of the mediation of two or more worlds that constitute an *Umwelt* (von Uexküll's [1982] definition of a "dwelling place" that arises from such interactions, presented previously in Chapter 4). The agent in this case refers to the new recording devices, such as autonomous recording units (ARUs), which digitally capture, store, and analyze sounds for the purposes of species identification. ARUs can record details as small as individual sonic emissions and as large as entire populations—sub-organismic, organismic, and super-organismic agencies (see Tønnessen, 2015).

Bioacoustics researchers today are primarily interested in method, technique, and representation due to the rapidly expanding datasets they have access to. While some use multiple ARUs to triangulate the position of organisms and their return to particular locations (which is the subject of Chapter 6), others use sound to measure how much anthrophonic factors cover natural sounds. An array of representational methods and knowledge syntheses are available to those interested in bioacoustics, moving well beyond the "manipulation and playback" model of acoustic ecology or the GIS-based representations of landscape ecology. Today, through simple apps and technologies that are easily deliverable and learnable though quick YouTube tutorials, data are cheaply and readily available for analysis through spectrogram or wave formats.

The Becoming Data of Sound

Again, Michael Gallagher (2015) refers to bioacoustics field recordings as "the nature style," where bioacoustics devices are capable of capturing vibrations imperceptible to normal human hearing (for instance, mice emit ultrasonic mating calls, plants emit infrasonic and ultrasonic vibrations in times of distress, and environments vibrate with the seismic activity of the earth). This recording technique attempts to erase the human perspective, despite the fact that human activity is intrinsically involved in locating environments untouched by humans (one must fly there, drive there, and step through plants to get there). Ironically, to remove the human-hearing element requires much human action and interaction.

The question of bioacoustics then is that of performance—the performance of nonhuman entities communicating with other nonhuman entities for the purposes of human capture and analysis. This is a long-standing practice in the Canadian context if we consider the work of R. Murray Schafer (1994) and his World Soundscape Project, begun in the 1960s, to collect ecological sounds. But bioacoustics research is based in the sensoria of the Anthropocene as a means of disrupting and changing our ethics and our morality, rather than simply as the aesthetic consideration of nature for its inherent qualities.

An interesting development here is the rising popularity of ultrasonic devices as a hobby in the UK, Canada, and the United States—such as apps or hardware that transduce sounds imperceptible to human hearing, like bat chirps, distress signals in rats, and botanical communication. Wildlife Acoustics (the maker of Song Sleuth), for example, sell handheld devices and smartphone extensions specifically for monitoring bats, bird and land animals, and marine life. Ultrasonic devices also transduce imperceptible vibrations into perceptible sounds in live time, allowing the recordist to analyze the frequency of bat movements. More expensive ultrasonic devices can often also record and slow down bat chirping. Measuring the beats and duration of wing sound, mid-flight body collapse, and echolocation chirping (which bats use to locate food sources) is intended for the UK's National Bat Monitoring Programme, to understand which bats are residing where, how their flight patterns are working, and where they are more and least active.

Even with the immensity of data accrued from citizen scientists, significant scepticism about its efficacy still exists in the scientific community. For instance, the eager, community-oriented, and naturalistic images of citizen scientists make some researchers wary about amateurs or voluntary biologists who claim to contribute to a further understanding of environmental issues (Greenwood, 2007). In contrast, an opposing argument suggests that as more people are drawn towards citizen science, newly defined components that legitimately belong to scientific knowledge emerge, rooted in the philosophy of the community-oriented approach to knowledge building, and the ethos of such citizen scientists groups.

However, for researchers, the term "big data" suffers from a "trough of disillusionment" (Sicular, 2013), as the capacity for data storage has reached from bit to byte to kilobyte to megabyte to gigabyte to terabyte to petabyte to exabyte to zettabyte to yottabite (or 2 to the power of 80 bytes, which Rob Kitchin [2014] describes as "too big to imagine" [p. 70]). Terms for even larger amounts will be necessary in the next decade. This disillusionment about the challenges of such massive data can be parsed by considering a couple of Latin antecedents: on one hand is *datum*, from which we get "data," that describes elements that have the potential to be extracted from phenomena; on the other is *captum*, which describes those elements which *are actually captured* from phenomena. The former is mere potential, the latter is actual.

Then and Now: Animal Recordings to Bioacoustics

It was Richard L. Garner's 1890s pioneering playback technique in Washington, DC, that inaugurated the interest in animal vocalizations (Bruyninckx, 2018). Garner recorded on a graphophone the sounds emitted by caged monkeys when they were poked with sticks by their handlers. He took particular interest in one monkey that "chattered" at its situation: a recording of these sounds were later played back to another monkey in the hopes of eliciting an artificial conversation and tapping into a "monkey language" (Radick, 2007). Garner linked his use of recording and playback technology to his beliefs that evolution offered insights into human experience. He also tied it to his view that sound technology exposed a language beneath the "lowest" forms of human life and deep within the evolutionary tract of primate life (Radick, 2007, p. 6). Bernard Stiegler (1998) has termed this union between technological advances and advancements in evolutionary thinking "epiphylogenesis," describing it as "the pursuit of the evolution of the living by other means than life" (p. 135).

The development of the spectrograph recorder across the mid-twentieth century helped to improve the study of animal vocalization through the visualization of sound, which inaugurated the field of bioacoustics (Mundy, 2009). Throughout much of the twentieth century, bioacoustics was used as a method for species identification, but the field is currently aligned with crisis-driven data collection and environmental monitoring. In sum, conceived as a broad-range monitoring tool, bioacoustics now has potential utility in a range of contexts beyond the original motivating monkey language:

1. It is a low-cost means of data collection: the transductors are possibly the cheapest of any technological monitoring devices on the market. In the face of decreasing funding opportunities for researchers, bioacoustics is not only reliable and accurate, but relatively inexpensive to supply to a research team (Servick, 2014).
2. It is far-reaching in scope, used to collect the sounds of plants, animals, humans, their communities, as well as human-made sound and noise. And, they can correlate human-made sounds with natural sounds (Allen, 2010).
3. The technologies have analytic tools built into them, so they can easily slow down recorded sounds, speed them up, analyze them automatically, register pace, frequency, intonation, amplitude, and silence (Tegeler, Morrison, & Szewczak, 2012).
4. It offers a means of collecting data without interference from the researcher, since they often will situate the devices and leave them for weeks or months at a time. Thus, human interference is minimized (Laiolo, 2010).
5. As bioacoustics devices transduce sounds digitally, virtually no long-term damage to the recordings themselves is possible, and they can be

uploaded immediately in live time to databases and research centres (Szostak, Gnoli, & López-Huertas, 2016).

Bioacoustics researchers have examined the role of sound and communication in the composition of an ecosystem, across a variety of organisms, and in rural, suburban, and urban habitats. Vertebrate and invertebrate animals, insects, and plants all emit sound to communicate and are capable of sensing changes in patterns of sound, which implies that communication between organisms is intentionally encoded by senders and intentionally decoded by receivers.

Birdsong has been of particular interest to bioacoustics researchers (Borker, Halbert, McKown, Tershy, & Croll, 2015; Kronenberg, 2014; Merchant et al., 2015; Rempel, Hobson, Holborn, Wilgenburg, & Elliott, 2005), in part because they change noticeably with the introduction of new stimulants (anthropogenic noise, for instance), but also because they are readily identifiable by the naked ear and are species-specific, so identification requires minimal professional training, if any. Birdsong is especially useful to monitor biodiversity loss because ecological gradients have effects on sensitive migratory acoustic communication patterns (Bayne, Habib, & Boutin, 2008). Again, bioacoustics in general, and birdsong in particular, reframes communication in such a way that includes human, nonhuman, and more-than-human, like sound-recording technologies into new non-hierarchical relationships.

How do research spaces open up through the use of new instruments and new contexts that bear an ethic of care? The closed lab culture of experts who isolate and observe animal behaviour—and who view the death and injury of animals as a generalizable sacrifice that serves a greater good (Lynch, 1988)—has given way to a more open, living lab made of empathic encounters and, especially, opportunities for new theorizing. Much of the expert modality has been problematized as an "epistemic scaffolding" that structures a myopically anthropocentric set of scientific conclusions (Friese & Clarke, 2012; Nelson, 2013), which reify isolated components of animal behaviour by stripping them of their social and historical contexts (Candea, 2013). But, as the laboratory opens up to public discourse, this includes folding public concerns into the "core sets" that guide research procedures (Michael & Birke, 1994), incorporating a co-emergence of research, researcher, and researched so that they hold comparable status within the research endeavour (Michael, 2012).

Accordingly, this opening has led to an interest in how laboratories condition encounters between human and nonhuman actors. Such examinations are shaped by the vocabulary and methodology of science and technology studies (Buller, 2013; Crist, 1996, 2004; Friese & Clarke, 2012), and the political discourses which inform how a "wild" animal becomes an object of knowledge (Asdal, 2008). On the other hand, some of this interest can be equally attributed to Donna Haraway's (2008) "contact zones," which has pushed for multispecies orientations towards knowledge production, and which lies beyond the closed model of research. This perspective underlies such recent empathy-centred concepts like "embodied empathy" (Despret, 2013), "trans-species empathy" (Chiew, 2014), and "care" (Giraud & Hollin, 2016).

So where does the field of bioacoustics intersect with the recent conversations around care and, in particular, "intimate sensing," described by Stefan Helmreich (2015) as a sensitivity to the multiplicity of an organism's sounds in the context of their environment? Turning to the terrain of sensing and intimate sensing will open bioacoustics to the larger interdisciplinary interest in technologies of sense, from the remote to the intimate. It is certainly obvious that microphones open onto the intimate dimensions of species production and reproduction. We must recall that bioacoustics is a widely interdisciplinary field of methods that record, store, and analyze human and nonhuman species sonifications; to generalize, bioacoustics is something that researchers *use* instead of study. It is a method whose interests lie in the soundings of bodies, and by extension how those soundings construct worlds that are sometimes conduits to others worlds, but sometimes barricades. But more so, it includes recordings produced by amateurs and professionals whose interests converge on biology and acoustics, broadly defined by the International Bioacoustics Council (2019) to include the following:

- animal sound production
- sound propagation in water and air
- insect vibrations
- biosonar and echolocation of bats and dolphins

- ultrasonic emissions
- infrasonic emissions
- hearing
- communication
- mating
- environments
- anthrosonic interference
- biodiversity monitoring
- sonic emissions for wildlife and pest control

It is thus largely a concrete field of study, with an eye to the empirics of sound production and analysis. Beyond this, however, is the sticky and complex intertwining of instrument, researcher, and researched. Bioacoustics researchers prefer sonic methods because sound equipment interferes far less with ecosystems and habitats, while still transmitting high-definition data. Despite this, they write often about the necessity to reduce the interference of researcher bodies with the researched bodies, as I wrote about in Chapter 4.

The astonishing range of sonic emissions of animals and other species has made it impossible for bioacousticians to speak of sound from an anthrocentric perspective. Generally speaking, bioacousticians consider sound as an ongoing and unfolding virtual mapping system for organisms to determine the proximity and size of other entities. The map is operationalized by the decibel count of a sound, which indicates the object's distance, and the frequency of the sound, which indicates the object's size. Humans, usually, cannot hear sounds that are above 20,000 hertz, nor below 20 hertz. Human ears also have decibel thresholds: a decibel that is too low cannot be perceived, and a decibel that is too high can do serious damage to human eardrums. As such, bioacoustics argue that humans live between sound worlds, only one of which they can perceive directly: sonic, infrasonic (below), and ultrasonic (high). Such a typecasting is narrow and limited for the study of animals in their environment. Instead, bioacousticians define animals by their acoustic traits—before their visual ones. They consider an animal's sound to be its very capacity to propagate and produce vibrational emissions and are interested in the ratio between wavelength and size.

Decentring Hearing, Recentring Listening

Bioacoustics has played a role in the development of a conservation biology as ornithologists have used the technology and techniques to identify species. This is a way to fortify variation, while reducing the noise input of phonographs and other technologies of sonic capture. On the other hand, newer conservation biologists prefer to use the technology to track biodiversity loss. When a species borders a high-impact anthropogenic zone on a regular basis, such as a highway or a factory, anthropogenic noise interferes with the volume and travel range of an organism's vocalization. Scientists can measure migration patterns against noise frequency and employ ARUs to capture ongoing recordings in real-time habitats, data they later analyze on a spectrographic visualization. Obviously, it would take innumerable hours to sit still and listen attentively, so researchers instead programme algorithms into small listening devices to isolate a certain range of sounds; they instal those devices into trees and close to nests, then retreat from the habitat altogether. This is a means to sonically map an ecosystem as well as to triangulate the location of species.

Preservation has long been a central cultural notion of technologies of modernity (photography, filmography, phonography, and so forth), but the realities generated by the technologies are complex. With emerging technologies, they also have available to them higher definition information, such as the nano-phones currently installed between the wings on medium-sized nocturnal avians like the common nighthawk discussed in Chapter 4 (see Rosen, 2017). Preservation is central to these new abilities, from preserving the microphones on the animals to preserving their sounds indefinitely.

Granular knowledge about animal populations and ecosystems is made possible due to an enormous repository of animals' sounds that are stored, visualized, and made widely available online. These visualized data simplify the process of identifying variation in acoustic streams, a practice previously reserved only for human ears trained to identify and catalogue species. In current bioacoustics, one "listens" but does so with eyes and fingers as well as ears. For instance, a research collaboration between Google Creative Lab, the Cornell Lab of Ornithology, and Cornell's Macauley

Library, called *Bird Sounds*, lays out thousands of birdsong sonogram samples onto a single digital map, organized according to song contours, tones, and rhythms. Certainly no map of any one place, the bird sounds peal whenever the cursor is dragged over a thumbnail sonogram impression. Resting the cursor on a sonogram impression triggers a splice of sound, but tracing the cursor over any cluster of them creates a sonic pixelated wave that sounds far from anything natural. The map is not intended to proffer any aesthetic contemplation of nature's symphony, which we might normally expect from an online database of bird sounds. Neither is it intended to map the places to which particular birds might belong, nor trace their migration patterns, which we traditionally expect maps to do. Instead, it is an algorithm sample board: an open-source code that makes possible the computer-generated classification of bird sounds by allowing the computer to listen to uploaded *.wav files, then automatically categorize them according to the contours of their vocalizations. The end result is a massive genealogical map, a tree with many offshoots.

The purpose of this interdisciplinary project is to creatively synthesize programming, conservationist intervention, species identification, biodiversity protection, and sustainability; it is intended to test the ability for a computer to make accurate classifications without the intervention of the human ear. The experiment is intended to open up sound, map, and place, and find new relationships between the social, collaborative, and spatial. This study includes sonic and creative mapping and spatial design—achieved through various sounding practices—so that the archive is a living, but impossible, ecosystem. Bird Sounds does not give us an actual ecosystem but constructs the semblance of one through a digital ecology, composed by programmers, researchers, algorithms, and the animals whose utterances are uploaded onto the programme. Bird Sounds is plainly a grid, which condenses and compresses song into splice, pushing on the borders of intelligibility.

Bioacoustics has become a defining skill set in the field of conservation biology (August et al., 2015). The technologies themselves have also undergone further development and improvement; they are now more powerful, have higher definition, and interact with the web and smartphones. They successfully facilitate wider participation and more creative inputs from interdisciplinary and transdisciplinary research teams, such as those using Bird Sounds. An especially notable improvement is their capacity to support non-invasive recording practices, especially in the context of contemporary environmental monitoring. According to Blumstein et al. (2011),

> with such technology, users can remotely and non-invasively survey human and animal populations, describe the soundscape, quantify anthropogenic noise, study species interactions, gain new insights into the social dynamics of sound-producing animals and track the effects of factors such as climate change and habitat fragmentation on phenology and biodiversity. (p. 758)

While earlier recordings in conservationist biology were intended to identify individual species, Hans Slabbekoorn and Magriet Peet's (2003) publication on birdsong and conservationist biology in *Nature* led to a surge of interest in the interactions between anthropogenic noise and biodiversity loss. This article also underscored the need for global research teams to contribute to the protection of species.

Positioning Listening Devices

It is the placement of the ARUs—more than the actual recording process—that is both embodied and haptic, as my accounts of the nighthawk monitoring in the last chapter taught me. This placement involves the co-presence of the researcher, the research tool, and the research subject. Digital technological developments have facilitated this recording ease, mainly through ARUs placed in field sites and left to record a longitudinal scope of sonic aggregation for weeks or months at a time. However, the locations of ARUs are often complicated, as they may be (1) accessible only by all-terrain vehicles, or (2) reachable only if the devices are lowered by helicopter or ridden in on fatbikes (an off-road bicycle equipped with oversized tyres to handle unstable terrain). Thus, it is in the *placement* of the ARU that we have an embodied corporeal presence, but those placements are at a high rate of anthropogenic interference. One researcher I interviewed noted this aspect:

The ARUs give us a lot of freedom to do long-term monitoring in inaccessible sites, but they can also make our field work more exciting. Not many of us have had the privilege of using the helicopter to drop an ARU, so we relish the challenge and experience when our supervisor suggests trying out new deployment methods. Dropping the ARU ... with a helicopter requires precision, because if the stand isn't placed carefully, it's gone. To pick up the ARU and stand, we have to go into the fen because the helicopter can't lift the stands without a technician there to attach it to the long-line. We have to go in to the edge of the fen by ATV or sometimes get dropped off, then hike our way in. We've only ever lost two units out of hundreds, and have amassed thousands of hours of data.

Researchers leave the ARUs on site to diminish their own anthropogenic interference on recordings, but this does not affect the interference of other factors—which is especially the case in so-called edge zones. These are sites where elements in one competing ecosystem will *mask* the capacity for communication in another, such as where a highway meets a forest. Masking refers to the level of anthropogenic sound that interrupt an animal's ability to vocalize, thus depleting the possibility for receivers to detect their calls and impeding on the possibility for mating and reproduction. Edge zones make it possible to determine whether species are capable of communicating over the mask of noise, like those from cars and factories. This research is critical to species' survival, for as Paola Laiolo (2010) writes, vocalization and the ability for organisms to hear are central to reproductive capacity:

> Just as genetic drift, bottlenecks and inbreeding can lead to a loss in genetic variation in small populations, cultural drift, bottlenecks and the reduced possibility of learning from models may determine the loss of acoustic diversity in species that learn their vocalizations. (p. 1640)

Another researcher I interviewed commented on this as well: "We consider *all* animals as defined by their acoustic parameters. Without a voice, they can't reproduce" (Vallee, personal communication, 2017).

Consideration of these factors moves bioacoustics beyond categorization by species (e.g. Obrist et al., 2010), since it takes into account the complexities of interspecies and intraspecies communication. It does so in a manner that is radically opposed to the orchestra of sounds, a characterization that has typified such appreciations, framing them as soundscapes—but bioacoustics recordings are transductions, as opposed to soundscapes, in that transductions catch so much more detail, adding both the infra- and ultrasonic dimensions to the recordings and mappings of communication patterns. Soundscapes have long been a point of contact for ecological listeners who use extensive technological apparatuses to capture the intertwining complexity of connections between embodiment, technology, and sites of nature.

From Sound Object to Sound Ecology

Among the many motivations for bioacoustics research is its response to the uncertainty and anxiety around biodiversity loss, on a global scale, and the role that anthrophonic interference is having on the balance of ecosystems. Some research teams, in fact, include "noise mapping" by reading city decibel levels that correspond to a colour-coded legend that identifies noise "hot-spots" (Hawkins, 2011). With approximately 83% of the land in the United States within two-thirds of a mile of a road, conservation officers team up with acousticians and sound ecologists to reduce the presence of helicopters, planes, and other means of transport into natural landscapes (Powers, 2016). Such high levels of noise have inspired Gordon Hempton to locate the "quietest square inch on earth" in the United States, ironically enough, claiming that it has no anthrophonic interference whatsoever for up to 20 minutes at a time (Berger, 2015).

Such efforts to locate quietude against the din of mobile humanity and expanding urbanization have also led researchers to take innovative conservationist measures, intended to "give back" the soundscape to the land (Berger, 2015). Specifically, they have selected habitats to geoengineer with sonic technologies—to coordinate new and better soundscapes by masking the anthrophonic interference with loudspeakers planted in natural settings. Others use multiple recording technologies to triangulate the exact location of species in an effort to expedite conservationist interventions for those animals that are deserting their natural habitat (Donaldson, 2016). *Triangulation* in this context enables the creation of a virtual space, where sound is used to trace the contours of what a place might become if intervention is not forthcoming. *Playback*, in contrast, is used for giving voice back to place.

Successful conservationist interventions must transcend local interventions and projects. This is the motivation behind films like *Global Soundscapes: Mission to Record the Earth*, an Imax feature that attempts to identify every sound in the world, towards the development of a massive online repository of the earth's sounds. Such interventions are intended to elucidate the making of a place through sound, but in such a manner that transcends any one such location. It is sonification that brings longitudinal research to life in ways that may actualize the global shifts referenced in the film title (Vartan, 2016). This is research that is geared towards instigating changes to "whole, global populations" of species, places, ecosystems, and the biosphere, driven by the mission to elucidate the necessity of conservationist intervention for the survival of the human race (Torino, 2015).

And to offset the unimaginably immense, local involvement of citizen scientists and volunteer biologists can contribute to the community-building aspects of global research initiatives. While the data collected and analyzed by these volunteers is sometimes perceived as borderline spurious (Cohn, 2008), the efforts for community building and for live feedback on scientific methodologies is invaluable. In Canada, students and community members are working with the University of British Columbia to sonically monitor tankers in the waters around Canada's proposed westbound pipeline (see Mazumder, 2016).

Certainly, if bioacoustics researchers are interested in whole populations over long periods of time, they must increasingly use technologies that listen to and recognize patterns—which include the declining sonic signals emitted by organisms. For example, to study "the whole spectrum of acoustical energy in a landscape" (Hall, 2016), soundscape ecologists in Germany have used 300 microphones to record one area; the microphones are timed to record one minute of sound in the environment every ten minutes, after which the data is processed by computer using over 120 terabytes of storage. This kind of sonic information is often fundamental for organisms' reproduction, but it is masked in places with more anthrophonic sonic drones, as shown in the 35,000 samples from recordings by one research team in Tippecanoe County in Indiana (Wallheimer, 2011). While organisms emit signals for particular purposes then, such as biological reproduction, the anthrophonic interference is an accidental byproduct of a machine in action. In theory at least, these sounds are meaning*less* beyond the action of the machine (Pijanowski, Farina, Gage, Dumyahn, & Krause, 2011; Pijanowski, Villanueva-Rivera et al., 2011), as opposed to the meaning*ful* sounds of the organisms.

As noted, soundscapes were initially introduced by Schafer (1994) and later edified by the World Soundscape Project at Simon Fraser University in Canada. Soundscapes are "alive by definition" (Blesser & Salter, 2007, p. 15). It is noteworthy that soundscape compositions have informed urban design (Adams et al., 2006; De Coensel & Botteldooren, 2007; Fong, 2016), cultural experience (Chandola, 2012), anthropogenic interference on natural habitats (Benschop, 2007), as well as waymaking and landmarks in spatial design (Chandrasekera, Yoon, & D'Souza, 2015). But soundscape *ecologies*—those oriented to the physiological and population health of a living organism or population in its ecosystem *or* in a laboratory setting—are more internal and intertwining, less intent on producing atmospheres than producing data about the endosonic and exosonic soundings of organisms. While field recordings remain part of the repertoire in bioacoustics (Farina, 2014; Farina, Lattanzi, Malavasi, Pieretti, & Piccioli, 2011; Farina, Pieretti, & Piccioli, 2011; Pijanowski, Farina et al., 2011; Pijanowski, Villanueva-Rivera et al., 2011), they proceed in a way that dramatically redefines the sound object of a soundscape, given a key characteristic of the ARU: its ability to capture an enormous range of longitudinal data. The variations contained within the data are perceptible by digital readers since the data are programmable through programmes such as Wildlife Acoustic's Kaleidoscope Analysis Software.

Soundscapes are used less within the context of soundscape ecology, where they are intended to raise awareness of environmental issues, and more within a conservation interventionist framework, where they are intended to provide data about anthropogenic influence on habitats. And bioacoustics researchers do not necessarily listen to sounds. In contrast to aesthetic contemplation, the scientific modality requires this data to be programmed into the machine algorithms that identify animals under investigation—especially acoustic animals like frogs and birds. One researcher described to the author the benefits of the ARU:

> There's no *way* to listen to that much data. There's no time for it either. I wouldn't characterize what we do as part of a "crisis discipline," but for conservationist efforts to be facilitated, we *have* to have all the data recorded and analyzed *quickly* using algorithms for identification. [...] That machines can listen for us now is remarkable. (Vallee, personal communication, 2017)

Thus, unlike the atmospheric or ambient listening that is typical of nature recordings or concentration mixes, bioacoustics research recordings are central to understanding the inner communication mechanisms of all forms of life. Indeed, conservation biologists study more than animals—they study botanical communication as well. For example, Ponomarenko et al. (2014) write on the recently discovered phenomenon of trees that scream in drought: "It has been established that ultrasonic emissions are more frequent under water stress conditions and are correlated with embolism patterns…suggesting that ultrasounds may be linked to cavitation" (p. 2). Generally speaking, this means that plants emit sonic information and are also receptive to surrounding sonic information.

When ultrasonic or infrasonic data are transduced, such a process introduces the impossibility of hearing any sense of original sound beyond human perception—there is always a sense of loss in transduction and in compression. "Lossy compression" was once a term used to designate the information lost during the data compression of audio and visual files (Manovich, 2001), and was approached cautiously in the early years of the digitization of sound. For many, the digitization of sound constituted a significant disconnect between the infrasonic and ultrasonic data that are usually present on many analogue recordings. For Jonathan Sterne (2003), for instance, this loss of deep resonance has made the digital sound file an arbiter in the passive listening paradigm, to revisit Eldritch Priest's (2018) recent denouncements. Lossy compression, however, had traded in the quality of a full range of sound for the space-saving files and devices many carry around with them.

In the context of this discussion, I am introducing the possibility of a "listening compression" as it grows from bioacoustics research, which simply describes how researchers rely on automata to identify data. Sense compression works in that it eliminates the necessity to listen in real time; it therefore deflects the act of listening, which includes resonance, understanding, and response. Lossy compression only poses a problem if we perceive the loss as an irrecoverable original, and if we discard the fact that all sounding is enmeshed in matters of transduction. Otherwise, every mode of transduction discards some information at the expense of other information, and if in this case the unnecessary information is cochlear stimulation in place of automation and visualization, listening might become immersed in the virtuand serve a longitudinal potential.

Concluding Remarks

To return once more to the orangutan from the previous chapter, emerging sound technologies open up new spatialized intimacies between researcher, researched, and the research process itself. In alignment with the turn towards species in cultural theory, I have aimed in this chapter to identify how animals' sounds, within such spatialized intimacies, are always meaningful. It is apparent that these spaces of meaning were opened through early sound-recording technologies (specifically, Garner's graphophone playback techniques [see Bruyninckx, 2018]), which clearly inaugurated a co-constitutive technosphere. Similarly, these new technologies made possible the perception of infra- and ultrasonic emissions otherwise inaccessible to human senses. Indeed, the spectrograph contributed significantly to the visualization of the "nether-sonic" worlds in animal utterances, which today are used to structure predications about biodiversity as well as to show how animals vocalize to respond to and manipulate their immediate environments. A shared ancestry has been theorized within this nether-sonic realm; it has also been the site for theorizations of new technologies, most relevant here being that of sound technologies.

Contemporary bioacoustics research expands outward to involve global research teams, with both professional and volunteer scientists, and includes dissemination that reassembles and reimagines global populations and ecosystems. It also codes listening technologies to map species according to their vocalizations. Demonstrably, long range analysis of animals' sounds in ecosystems requires new listening tools for storing an expanded duration of digital data. The pragmatic uses of emerging sound technologies are obvious: early detection of biodiversity loss, faster rates of conservationist interventions, a deeper sense of shared ancestry. But much of these benefits arrive at the expense of de-anthropocentrizing the listening experience. This goes against the grain of cultural theories of sound and sounding, which themselves privilege the experiential, the perspectival, and the aesthetic configurations of sonic immersion. Indeed, if a tree falls in the forest, it should be for more than a philosopher to wonder whether it makes a sound.

Emerging technologies thus decentre anthropocentric hearing, but they recentre the relations that humans might have with other species, given that the sense of hearing expands beyond that which any human can hear unaided by technology. This process thereby generates the possibility for connections to that to which we were previously insensible. The scientific perspective shows us how we hear just a slice of the sonic spectrum; further, sound is a method for elucidating how researchers are implicated in their research, including the animal–human–sound–species–techno–listening worlds that contain and expand their capacities and potentials.

References

Adams, M., Cox, T., Moore, G., Croxford, B., Refaee, M., & Sharples, S. (2006). Sustainable soundscapes: Noise policy and the urban experience. *Urban Studies, 43*(13), 2385–2398.

Allen, M. (2010). VoxNet: Reducing latency in high data rate applications. In E. Gaura, L. Girod, J. Brusey, M. Allen, & G. Challen (Eds.), *Wireless sensor networks: Deployments and design frameworks* (pp. 115–158). London, UK: Springer.

Asdal, K. (2008). Subjected to parliament: The laboratory of experimental medicine and the animal body. *Social Studies of Science, 38*(6), 899–917.

August, T., Harvey, M., Lightfoot, P., Kilbey, D., Papadopoulos, T., & Jepson, P. (2015). Emerging technologies for biological recording. *Biological Journal of the Linnean Societ, 115*(3), 731–749.

Back, L. (2007). *The art of listening.* London, UK: Berg.

Baptista, L. F., & Keister, R. A. (2005). Why birdsong is sometimes like music. *Perspectives in Biology and Medicine, 48*(3), 426–443.

Bayne, E. M., Habib, L., & Boutin, S. (2008). Impacts of chronic anthropogenic noise from energy sector activity on abundance of songbirds in the boreal forest. *Conservation Biology, 22*(5), 1186–1193.

Benschop, R. (2007). Memory machines or musical instruments? Soundscapes, recording technologies and reference. *International Journal of Cultural Studies, 10*(4), 485–502.

Berger, E. (2015). Welcome to the quietest square inch in the U.S. outside. *Outside Online.* Retrieved from https://www.outsideonline.com/2000721/welcome-quietest-square-inch-us.

Blesser, B., & Salter, L. R. (2007). *Spaces speak, are you listening? Experiencing aural architecture.* Cambridge: MIT Press.

Blumstein, D. T., Mennill, D. J., Clemins, P., Girod, L., Yao, K., Patricelli, G., … Kirschel, A. (2011). Acoustic monitoring in terrestrial environments using microphone arrays: Applications, technological considerations and prospectus. *Journal of Applied Ecology, 48*(3), 758–767.

Borker, A. L., Halbert, P., McKown, M. W., Tershy, B. R., & Croll, D. A. (2015). A comparison of automated and traditional monitoring techniques for marbled murrelets using passive acoustic sensors. *Wildlife Society Bulletin, 39*(4), 813–818.

Bourdieu, P. (1984). *Distinction: A social critique of the judgement of taste.* Cambridge, MA: Harvard University Press.

Brand, A. R. (1937). Why bird song cannot be described adequately. *The Wilson Bulletin, 49*(1), 11–14.

Bruyninckx, J. (2018). *Listening in the field: Recording and the science of birdsong.* Cambridge: MIT Press.

Buller, H. (2013). Individuation, the mass and farm animals. *Theory, Culture & Society, 30*(7–8), 155–175.

Candea, M. (2013). Habituating meerkats and redescribing animal behaviour science. *Theory, Culture & Society, 30*(7–8), 105–128.

Chandola, T. (2012). Listening in to water routes: Soundscapes as cultural systems. *International Journal of Cultural Studies, 16*(1), 55–69.

Chandrasekera, T., Yoon, S.-Y., & D'Souza, N. (2015). Virtual environments with soundscapes: A study on immersion and effects of spatial abilities. *Environment and Planning B: Planning and Design, 42,* 1003–1019.

Chiew, F. (2014). Posthuman ethics with Cary Wolfe and Karen Barad: Animal compassion as trans-species entanglement. *Theory, Culture & Society, 31*(4), 51–69.

Cohn, J. P. (2008). Citizen science: Can volunteers do real research? *AIBS Bulletin, 58*(3), 192–197.

Connors, J. P., Patrick, J., Lei, S., & Kelly, M. (2012). Citizen science in the age of neogeography: Utilizing volunteered geographic information for environmental monitoring. *Annals of the Association of American Geographers, 102*(6), 1267–1289.

Cornwell, M. L., & Campbell, L. M. (2012). Co-producing conservation and knowledge: Citizen-based sea turtle monitoring in North Carolina, USA. *Social Studies of Science, 42*(1), 101–120.

Crist, E. (1996). Naturalists' portrayals of animal life: Engaging the verstehen approach. *Social Studies of Science, 26*(4), 799–838.

Crist, E. (2004). Can an insect speak? The case of the honeybee dance language. *Social Studies of Science, 34*(1), 7–43.

De Coensel, B., & Botteldooren, D. (2007). The rhythm of the urban soundscape. *Noise and Vibration Worldwide, 38*(9), 11–17.

Despret, V. (2013). Responding bodies and partial affinities in human–animal worlds. *Theory, Culture & Society, 30*(7/8), 66–91.

Donaldson, A. (2016, July 11). National network of acoustic recorders proposed to eavesdrop on australian ecosystems. *ABC News.* Retrieved from http://www.abc.net.au/news/2016-07-11/soundscape-ecology-could-track-environmental-changes/7587354.

Farina, A. (2014). *Soundscape ecology: Principles, patterns, methods and applications.* London, UK: Springer.

Farina, A., Lattanzi, E., Malavasi, R., Pieretti, N., & Piccioli, L. (2011). Avian soundscapes and cognitive landscapes: Theory, application and ecological perspectives. *Landscape Ecology, 26*(9), 1257–1267.

Farina, A., Pieretti, N., & Piccioli, L. (2011). The soundscape methodology for long-term bird monitoring: A Mediterranean Europe case-study. *Ecological Informatics, 6*(6), 354–363.

Fong, J. (2016). Making operative concepts from Murray Schafer's soundscapes typology: A qualitative and comparative analysis of noise pollution in Bangkok, Thailand, and Los Angeles, California. *Urban Studies, 53*(1), 173–192.

Friese, C., & Clarke, A. E. (2012). Transposing bodies of knowledge and technique: Animal models at work in reproductive sciences. *Social Studies of Science, 42*(1), 31–52.

Gabrys, J. (2016). *Program earth: Environmental sensing technology and the making of a computational planet.* Minneapolis: University of Minnesota Press.

Gallagher, M. (2015). Field recording and the sounding of spaces. *Environment and Planning D: Society and Space, 33*(3), 560–576.

Giraud, E., & Hollin, G. (2016). Care, laboratory beagles and affective utopia. *Theory, Culture and Society, 33*(4), 27–49.

Greenwood, J. J. D. (2007). Citizens, science and bird conservation. *Journal of Ornithology, 148*(Suppl. 1), S77–S124.

Hall, M. (2016). Soundscape ecology: Eavesdropping on nature. *Deutsche Well (DW).* Retrieved from http://www.dw.com/en/soundscape-ecology-eavesdropping-on-nature/a-19304871.

Haraway, D. (2008). *When species meet.* Minneapolis: University of Minnesota Press.

Hawkins, D. (2011). "Soundscape ecology": The new science helping identify ecosystems at risk. *Ecologist: Setting the environmental agenda since 1970.* Retrieved from https://theecologist.org/investigations/science_and_technology/1171165/soundscape_ecology_the_new_science_helping_identify_ecosystems_at_risk.html.%20Accessed%20September%2030,%202017.

Helmreich, S. (2015). *Sounding the limits of life: Essays in the anthropology of biology and beyond.* Princeton: Princeton University Press.

Hill, P. (2007). *Olivier messiaen: Oiseaux exotiques.* Burlington, VT: Ashgate.

International Bioacoustics Council. (2019). http://www.ibac.info.

Irwin, A. (1995). *Citizen science: A study of people, expertise and sustainable development.* London: Routledge.

Johnson, J. J. (1995). *Listening in Paris: A cultural history.* Berkeley, CA: University of California Press.

Kircher, A. (1650/1970). *Musurgia universalis.* Hildesheim and New York: G. Olms (Reprint of the Rome, 1650 edition).

Kitchin, R. (2014). *The data revolution: Big data, open data, data infrastructures, and their consequences.* London, UK: Sage.

Kronenberg, J. (2014). Environmental impacts of the use of ecosystem services: Case study of birdwatching. *Environmental Management, 54*(3), 617–630.

Laiolo, P. (2010). The emerging significance of bioacoustics in animal species conservation. *Biological Conservation, 143*(7), 1635–1645.

Latimer, J. (2013). Being alongside: Rethinking relations among different kinds. *Theory, Culture & Society, 30*(7/8), 77–104.

Latour, B. (2005). *Reassembling the social: An introduction to actor-network-theory.* Oxford: Oxford University Press.

Lawrence, A. (2006). "No personal motive?" Volunteers, biodiversity, and the false dichotomies of participation. *Ethics, Place and Environment, 9*(3), 279–298.

Lievrouw, L. A. (2010). Social media and the production of knowledge: A return to little science? *Social Epistemology, 24*(3), 219–237.

Lynch, M. E. (1988). Sacrifice and the transformation of the animal body into a scientific object: Laboratory culture and ritual practice in the neurosciences. *Social Studies of Science, 18*(2), 265–289.

Manovich, L. (2001). *The language of new media.* Cambridge: MIT Press.

Mazumder, A. (2016). *Pacific North West LNG Project: A review and assessment of the project plans and their potential impacts on marine fish and fish habitat in the Skeena estuary.* Environmental Assessment Report, Government of Canada. Minister of Environment and Climate Change. https://www.ceaa.gc.ca/050/evaluations/proj/80032?culture=en-CA.

Merchant, N. D., Fristrup, K. M., Johnson, M. P., Tyack, P. L., Witt, M. J., Blondel, P., & Parks, S. E. (2015). Methods in ecology and evolution. *Methods in Ecology and Evolution, 6*(3), 257–265.

Michael, M. (2012). Anecdote. In C. Lury & N. Wakeford (Eds.), *Inventive methods: The happening of the social* (pp. 25–35). London, UK: Routledge.

Michael, M., & Birke, L. (1994). Enrolling the core set: The case of the animal experimentation controversy. *Social Studies of Science, 24*(1), 81–95.

Morton, T. (2010). *The ecological thought.* Cambridge, MA: Harvard University Press.

Mundy, R. (2009). Birdsong and the image of evolution. *Society and Animals, 17,* 206–223.

Nelson, N. C. (2013). Modeling mouse, human, and discipline: Epistemic scaffolds in animal behavior genetics. *Social Studies of Science, 43*(1), 3–29.

Obrist, M. K., Pavan, G., Sueur, J., Riede, K., Llusia, D., & Marquez, R. (2010). Bioacoustics approaches in biodiversity inventories. In J. Eymann, J. Degreef, C. L. Häuser, J. C. Monje, Y. Samyn, & D. Vandan Spiegel (Eds.), *Manual on field recording techniques and protocols for all taxa biodiversity* inventories (pp. 68–99). Brussels: Belgian Development Cooperation.

Pijanowski, B. C., Farina, A, Gage, S. H., Dumyahn, S. L., & Krause, B. L. (2011). What is soundscape ecology? An introduction and overview of an emerging new science. *Landscape Ecology, 26*(9), 1213–1232.

Pijanowski, B. C., Villanueva-Rivera, L. J., Dumyahn, S. L., Farina, A., Krause, B. L., Napoletano, B. M., & Pieretti, N. (2011). Soundscape ecology: The science of sound in the landscape. *BioScience, 61*(3), 203–216.

Ponomarenko, A., Vincent, O., Pietriga, A., Cochard, H., Badel, É., & Marmottant, P. (2014). Ultrasonic emissions reveal individual cavitation bubbles in water-stressed wood. *Journal of the Royal Society Interface, 11*(99), 20140480.

Powers, A. (2016). Preserving the quietest places. *The California Sunday Magazine.* Retrieved from https://story.californiasunday.com/quietest-places-on-earth.

Priest, E. (2018). Earworms, daydreams and cognitive capitalism. *Theory, Culture & Society, 35*(1), 141–162.

Radick, G. (2007). *The simian tongue: The long debate about animal language.* Chicago, IL: Chicago University Press.

Rempel, R. S., Hobson, K. A., Holborn, G., Wilgenburg, S. L. V., & Elliott, J. (2005). Bioacoustic monitoring of forest songbirds: Interpreter variability and effects of configuration and digital processing methods in the laboratory. *Journal of Field Ornithology, 76*(1), 1–11.

Rosen, J. (2017). Sustainability: A greener future. *Nature, 546*(7659), 565–567.

Rothenberg, D. (2008). *Thousand-mile song: Whale music in a sea of sound.* New York, NY: Basic Books.

Schafer, R. M. (1994). *The soundscape.* Rochester, VT: Destiny Books.

Servick, K. (2014). Eavesdropping on ecosystems. *Science, 343*(6173), 834–837.

Sicular, S. (2013, January 22). *Big Data is falling into the trough of disillusionment.* Retrieved from Gartner database. https://blogs.gartner.com/svetlana-sicular/big-data-is-falling-into-the-trough-of-disillusionment/.

Slabbekoorn, H., & Peet, M. (2003). Birds sing at a higher pitch in urban noise. *Nature, 424*, 267.

Sterne, J. (2003). *The audible past: Cultural origins of sound reproduction.* Durham, NC: Duke University Press.

Stiegler, B. (1998). *Technics & time 1: The fault of epimetheus* (R. Beardsworth, Trans.). Stanford, USA: Stanford University Press.

Szostak, R., Gnoli, C., & López-Huertas, M. (2016). *Interdisciplinary knowledge organization.* New York: Springer International Publishing.

Tegeler, A. K., Morrison, M. L., & Szewczak, J. M. (2012). Using extended-duration audio recordings to survey avian species. *Wildlife Society Bulletin, 36*(1), 21–29.

Tønnessen, M. (2015). The biosemiotic glossary project: Agent, agency. *Biosemiotics, 8*, 125–143.

Torino, L. (2015). You can actually hear the climate changing. *Outside.* Retrieved from https://www.outsideonline.com/2035701/you-can-actually-hear-climate-changing.

Truax, B. (2001). *Acoustic communication* (Vol. 1). Westport: Greenwood Publishing Group.

Vallee, M. (2017). The rhythm of echoes and echoes of violence. *Theory, Culture & Society, 34*(1), 97–114.

Vartan, S. (2016). We're changing the way the world sounds: Noise impacts ecosystems in more ways than you might think. *Mother Nature Network*. Retrieved from http://www.mnn.com/earth-matters/wilderness-resources/blogs/we-are-changing-way-world-sounds.

von Uexküll, J. (1982). Glossary. *Semiotica, 42*(1), 83–87.

Wallheimer, B. (2011, March 23). New scientific study will study ecological importance of sounds. *Science News*. Retrieved from https://www.sciencedaily.com/releases/2011/03/110301122154.htm.

6

Sounding Place

This last empirical chapter experiments with the sounding of place, analyzing a rhythm of disaster, as I attempt to articulate an infrastructure of audibility beneath the tourist industry, one that becomes incorporated into the making of disaster. The chapter surveys the historical and contemporary geoscientific assessment of the Frank Slide, well-known as Canada's "Deadliest Rockslide." I follow this with an examination of the rhythm of cultural work that is folded into the tourist industry, fictionalization, and imperceptibility, to fully explore what sounding might accomplish as a method if it were used in a context where sound is not necessarily a binding sense. Recall, for instance, that the infrastructure of audibility refers less to the notion of listening and more to the notion of audibility—to resonating, of being in rhythm, of timing, and place. My analysis considers how these disparate activities contribute to the vitalization, devitalization, and revitalization of place, in such a way that challenges the overly simplified "dark tourism" hypothesis that has come to frame disaster sites (Bowman & Pezzullo, 2010). The chapter thus poses a robust dual treatment of the concept of sounding: it can be applied to non-sonorous forces, but also contributes to the study of place, time, and disaster. This segment will close with a mention of the resonant temporalities, speeds, and time horizons

© The Author(s) 2020
M. Vallee, *Sounding Bodies Sounding Worlds*, Palgrave Studies in Sound,
https://doi.org/10.1007/978-981-32-9327-4_6

that go into the making of place, specifically with a discussion of Doreen Massey's concept of event of place and Timothy Morton's (2013) complementary concept of the very large finitude. Thus, while I approach the concept of sound somewhat differently in this chapter, sounding remains the persistent theme here. But to begin, I turn first to the conceptualization of rhythm and how we might understand it within a cultural studies framework.

Conceptualizing Rhythm: Mileux and Transformations

How is place an imaginary organ? And how is it distributed through its infrastructure of audibility over larger timescales than even computational bioacoustics can grapple with? This chapter examines the relationships between rhythm, repetition, and the politics of disaster in a tourist destination in Southern Alberta, Canada. Moving beyond the bioacoustic and into the vibrations of the earth, I argue in this chapter that such vibrational moves *cannot* be sonified in any manner and must be made sense of through abstractions of place. Rhythm is a kind of audibility that is more felt than it is heard—as such, it is an abstraction that will be a central focus here, as is the successive historical conception of rhythm. Indeed, rhythm can be useful for accessing multiple time frames and temporalities; at the very least it is an interlocking mechanism between the movements of daily life and the broader social milieux that give shape to those movements. Rhythm thus underlies the mechanistic production of place.

This chapter contributes further to the infrastructure of audibility by way of a cultural analysis of rhythm in time and place, which I use as a methodology to theorize an imaginary organ of place through an instrumental case study. Exploring the Frank Slide rockslide in Southern Alberta Canada offers an expanded and expanding sense and sensation that embraces rhythm and sounding. Rhythmanalysis—as an ambivalent and contradictory heuristic device—will ground this analysis and allow entry into new relations between time, repeatability, and possibility.

Rhythm is generally conceptualized in cultural theory as a disrupting but reinforcing force that shapes a subject. However, I will approach rhythm in a material-historical mode, insofar as rhythm proliferates in varying material, immaterial, abstract, and nascent forms. The chapter thus contributes to sound studies, but in a renewed sociological sense of social and cultural history that emphasizes the simultaneous over the successive, the cyclic over the linear, and the rhythmic over repetition. These aspects are made manifest via encounters, extensions, becomings, and contradictions, and their destinations towards new calculations of social transformation. To expand, the following specific theorization of *rhythm as violence* is drawn from Gilles Deleuze and Félix Guattari (1987), pivoting around the themes of *breaching, reaching,* and *returning.* These themes constitute the infrastructure of audibility for rhythm, or how an event and a place are produced through multiscalar and multidimensional timescales and time horizons, experienced simultaneously.

In *A Thousand Plateaus,* Deleuze and Guattari use rhythm in an unconventional manner, just as they similarly presented a unique perspective of voice as the site for the incorporeal transformation of subjectivity. Rhythm here has less to do with the flow of time than time's utterly violent reorganization through breaches, reaches, and returns. First, rhythm breaches. "There is nothing less rhythmic than a military march," Deleuze and Guattari declare in *A Thousand Plateaus* (1987, p. 345). In a way, this is true, even while it may not necessarily be apparent initially. When soldiers march over a bridge, for instance, they must "break stride" to prevent the bridge from rupturing under the cadence of their unified steps. The consequence of the military march—amplitude—is indeed rhythmic according to Deleuze and Guattari's conception. For example, when milieux encounter one another—here, the milieu of the military march against the milieu of the bridge's structure—each milieu produces in the other a transformation towards a new milieu. A disaster-milieu could be a bridge disaster like the Broughton Suspension Bridge of 1831 or the Angers Bridge of 1850; an innovation-milieu might be structures that are capable of withstanding the stress of amplitude.

Deleuze and Guattari's conception of rhythm turns into a point of rupture and mutual transformation between the interpenetrating milieux of military march and bridge structure. Now, any milieu or refrain can participate in this reciprocal exchange, so that the defining feature of rhythm is its unexpected transformation, arising from the interaction between milieux.

Simply put, Deleuze and Guattari's theory of rhythm accounts for how a collection of disparate milieux can together become a new, broader milieu. Milieux are primarily transformational despite their tendency to be metrically bound by a persistent repetition, such as the repetition of steps in a military march; further, they continuously merge with new times and new spaces. The consequence of intermingling milieux is a rhythm or an unforeseen accent in repeated patterns whose amplitude is the maximum measure of power between the milieux. Thus, in isolation, the military march, or any milieu for that matter, is not truly rhythmic. It must enter into an arrangement with an unforeseen other milieu, or both milieux must enter into a transduction of pure transformation. In the case of the collapsing bridge, that transduction is chaos, which Deleuze and Guattari consider to be the closest filiation to rhythm.

Deleuze and Guattari's (1987) claim regarding rhythm focuses on the relation between milieux. Taking a quasi-embryonic perspective, Deleuze and Guattari conceive of political and social organizations as spacetimes between, around, and within milieux. These are made of material, substance, membranes, energy, and perceptions. For them, milieux are fundamentally coded through a regular pulse, a repetition. The code is continuously subjected to the laws of transduction, however, in that signals are continuously converted into other signals, in a process that is intended to thwart chaos: "The milieux are open to chaos, which threatens them with exhaustion or intrusion" (1987, p. 345). Rhythm is the key to thwarting chaos, which itself does not stand outside any assemblage of milieux but is instead "the milieu of all milieus" (p. 345)—not a noise in space, but a space opened up by noise. That is why a military march alone is not rhythmic: it does not pass from milieu to milieu, but is itself a milieu coded through the repetition of a refrain:

> Meter, whether regular or not, assumes a coded form whose unit of measure may vary, but in a noncommunicating milieu, whereas rhythm is the Unequal or the Incommensurable that is always undergoing transcoding. Meter is dogmatic, but rhythm is critical...[so that it] ties together critical moments, or ties itself together in passing from one milieu to another. (p. 346)

While it is not uncommon to think of rhythm as a flow or a groove, stable and unchanging, rhythm for Deleuze and Guattari (1987) represents, in contrast, a force of local transformation and is thus unstable, transformative, and violent. If rhythm is our focus, then, it is not enough to designate it to any one specific process, but rather to understand it as a continuum. Rhythm, Deleuze (2003) writes in his book on Francis Bacon, invests itself in levels of the senses, which change variously across different milieux. A Deleuzean reading of rhythm concentrates on the changes, which manifest as violent ruptures of difference that might appear as chaotic (in the song, the figure, the rebuilding), but that are followed by regularity in the return to the refrain, to repetition.

There is a profound variation in the media through which rhythm is perceived: we might feel rhythm in dance, hear rhythm in music, see rhythm in painting, smell rhythm in preparing food, and taste rhythm in our consumption of it. Rhythm links the universal, as rhythm takes place across all vibrations and is intrinsic to embodied experience. Elizabeth Grosz (2008) writes that all cultural media correspond with the production of place, while also preparing the potential destruction of that place, arguing that rhythm erupts from people's embodied encounters with vibratory media, which is in turn implicated in the production of place. These offer only a momentarily stable situation before everything is undone and undergoes another transformation. A milieu resonates with rhythm, making territory and quality part of the same movement.

Rhythm, however, is more complex than articulating an embodied transformation. To extend such a conceptualization, rhythm is irrational because it "can only be looked at and experienced" (Bode, 2014, p. 55). It makes allowance for the habits that shape a body's internalized social values and norms; rhythm is a crossover of internal and external realities. As such, it is "quite typical of how bodies work" (Henriques, 2014, p. 104). Obvious musical implications reside here: rhythm is an auditory and embodied phenomenon that arises from a pattern of bodies striking one another. For instance, when different musicians play patterns on instruments, they enter into *a* collective rhythm, but they do not play "individual rhythms." Rhythm is something of a virtual object that actors contribute to through processions, which themselves become objects that others respond to in a rhythm network, involving the striking, plucking, or rubbing of objects together that are designed (usually) to emit sound.

This is possibly why Henri Lefebvre's (2004) rhythmanalysis has proven so influential, since he embraces the culturally embodied constitution of rhythm without losing sight of its technical detail. Lefebvre admits that rhythm is elusive but maintains that its measurements are palpable. A rhythmanalyst can easily move between milieux and measurements and connect them in way that uncovers their overarching rhythm. Rhythm-analysis is a method that searches for the general production of "freedoms" (Lefebvre, 2004, p. 32) that account for the auditory mechanics of social transformation. That is, if repetition is central to rhythm, which it is, then it cannot be articulated meaningfully without the articulation of difference.

Rhythm has itself germinated a fascinating array of case studies, ranging from the "rhythms" of social institutions such as museums (Prior, 2011), to time-lapse photography (Simpson, 2012), soundwalking (Hall, Lashua, & Coffey, 2008), and mobility studies generally (DeLyser & Sui, 2013). In much of this work, rhythm unintentionally becomes an interchangeable conceptual category with tempo and speed, or even more generally, with life and vitality. In resonance with Michael E. Gardiner's (2012) obser-vation that Lefebvre holds a tenuous and non-systematic commitment to the concepts he introduces, rhythmanalysis's mechanics of rhythms (like tempo, speed, and duration) are thus equivocated, nonspecific, and ambivalent. The rhythmanalyst of postmodernity by consequence seems no more removed from self-reflexive observation than the clichéd flâneur of modernity (see Shields, 2006). In recent years, as cultural theories of rhythm have both emerged and resurfaced, scholars have adopted and modified rhythm to suit a wide range of macro-, meso-, and micro-level studies, which leads to a spectrum of studies from the rhythmic transformations of history (Gardiner, 2012) to the pulses of everyday life (Simpson, 2012). Indeed, the interlacing of smaller and larger pulses and patterns constitutes a central defining feature of rhythm.

What follows is a rhythmanalysis of place, to elucidate how rhythm is an infrastructure of audibility that we can use in research to produce the multiple timescales of place. This is thus the least sonically determined chapter in the book; instead, it is devoted to the geographical timescales of finitude and disaster.

Imperceptible Milieux

Spreading out east of the continental divide, north of the Montana/Alberta border, stretched over the Crowsnest Highway (Highway 3), Google Maps satellite imagery reveals a small grey smudge of rubble and debris. The Frank Slide, "Canada's deadliest rockslide," scars the digital imagery. When viewed in person, it is as sublime as it is ordinary, dividing the Municipality of Crowsnest Pass, a quiet coal-mining consolidation of villages and towns with a growing population. Having collapsed on the town of Frank on April 29, 1903, at 4:10 a.m., killing between 70 and 90 residents, the Slide is now a provincial heritage site. Of course, in keeping with memorial sites and place-based memorial site literature (Degnen, 2005), it is a place-bound memory, but it is an easy-access tourist destination, a boulder-climber's mecca, a geologist's fascination, a pika's habitat. For cultural researchers (Ang, 2006), it is a place made of heterogeneous times and temporalities. Here, it is also a case study for a new "edge method" in social scientific research that traces the contours of representational frameworks.

The mountain that loosened the slide is known as "Turtle Mountain," named as such because it moves incrementally every day (Benko & Stead, 1998); "the mountain that moves" constitutes part of the "place-image" (Shields, 1991) that runs from First Nations knowledge to contemporary schoolroom education (Marty, 2011). The Frank Slide is also known to geologists as just partially collapsed, a prelude to more destruction, which could occur right now or anytime over the next 5000 years, and is monitored by seismic readers that respond to the slightest of movements. The Frank Slide Interpretive Centre, a historic resources management site, is where activity from the Slide's monitoring stations is recorded and sent to research stations throughout Alberta, as well as to an international research station in Italy. Tourists are welcome to view the Slide through several telescope stations on a walking path around the centre. The centre's website has both an assurance and a warning:

Scientists estimate that any major rockslide from Turtle Mountain is not likely to occur anytime soon if the mountain continues to move at its current slow, turtle-like pace. However, if movement begins to speed up, a major rockslide could happen much sooner. And, if an earthquake happens nearby, all bets are off. (Alberta Culture and Tourism, 2016, https://frankslide.ca/learn)

Consistent with research conducted at other disaster sites, the Frank Slide is monitored using geological sensors that respond to the otherwise imperceptible vibrations. NavStar Geomatics Ltd. installed the following surveillance:

- 8 differential global positioning satellites that signal movements based on their relational position with one another
- 14 periodic global positioning site readings at two particularly fragile locations (the "saddle zone" and between the third and south peaks at the eastern face of the mountain)
- 1 Ground-Based Interferometric Synthetic Aperture Radar (ISAR) to scan the scar of the Slide every 8 minutes on a continuous cycle (Alberta Energy Regulator, 2016c).

I am told by some of the residents in my fieldwork that this final item of technology is the only operative technology now, since the others are broken due to regular wear and human interference (Vallee, 2018).

Rhythm is a useful interdisciplinary framework to examine this historic event and site, since there are many other rhythms (aside from naturally occurring ones) that must be taken into account when theorizing a place like the Frank Slide: it is a tourist site, a provincial heritage site, and a subject of popular culture. The Frank Slide has other significant attributes as well: it is the site of Alberta's most diverse bird population, surrounded by some of the best river trout fishing in Western Canada; it is situated in Alberta's only historically known 1930s Communist-led municipality (Harrison & Friesen, 2015); and it is the fascination of journalists, scientists, artists, and people who "boulder." As such, it embodies multiple times, speeds, and temporalities. And, I contend, these times and temporalities are constituent participants whose intersections with one another contribute towards the production of place.

Measurement Milieux

In 1903 a catastrophic rock slide at the North Peak of Turtle Mountain buried parts of the town of Frank, killing between 70 and 90 people. Since that time extensive geological studies have been conducted to determine the likelihood of another slide, and have reached the conclusion that the potential exists for another devastating slide from the North, South, or Third Peak of Turtle Mountain. (Government of Alberta, 2017, p. 5)

To monitor a disaster site successfully, as the literature on disaster preparedness demonstrates (Hernández-Moreno & Alcántara-Ayala, 2016; Sassa, 2016; Sassa & Canuti, 2008; Wachinger, Renn, Begg, & Kuhlicke, 2013), it needs continuous scientific monitoring along with efficient communication between scientists and the public, and evacuation plans. The development of automated technological monitoring systems has thus improved the prediction rate of landslides (Stähli et al., 2015), though, historically, researchers have creatively used an array of instruments. Monte Arbino (Switzerland), for instance, was monitored continuously using a trigonometric network, involving routine surveying from 1889 to 1928; the land finally fell on October 2, 1928 and 14 households were successfully evacuated (Bonnard, 2011). Routine photography of a potential disaster site also produced reliable monitoring results, such as the post-slide photographs of Elm in Switzerland (Melosh, 1986). After the Frank Slide, volunteers monitored the fissures at the south peak created by the disaster until 1931, when a government-funded research project isolated a "danger zone" and ordered residents in that zone to move from the path of the next slide (Wilson, 2005).

After its collapse, influential studies concluded that the disaster was the consequence of bad mining practices (McConnell & Brock, 1904; Sharpe, 1938; Terzaghi, 1961), attributed largely, though far from exclusively, to the Canadian American Coal and Coke Company mine in operation since 1900, situated at the base of the mountain. However, mining has since been reduced to a negligible contributing factor (Cruden & Martin, 2007). Turtle Mountain was long unstable, a mountain made of limestone overtop of shale, with an uneasy base attributed to a vast overhang. The Frank Slide, to offer an accurate but grossly simplified account, is the result of unstable and incompatible edges converging with one another. Geologist Karl Terzaghi (1950) wrote of this:

> In hard, jointed rocks resting on softer rocks…a decrease of the cohesion of the rock adjoining a slab may occur on account of creep of the softer rocks forming their base… the limestones, forming the bulk of the peak, rested on weaker strata which certainly crept under the influence of the unbalanced pressure produced by the weight of the limestone and *the rate of creep was accelerated by coal-mining operations in the weaker strata.* (pp. 95–96, emphasis added)

The Frank Slide was by no means the first mountain landslide perceived as being *partly* caused by industry. The 1881 "Rockslide of Elm" at the Swiss Canton of Glarus was attributed to bad mining practices, for instance (Stahr & Langenscheidt, 2014). There was no sense among the miners in the town of Frank that the "mountain that moves" would collapse due to the placement of the mine. In fact, research now shows the mine was strategically placed (Cruden & Martin, 2007). Ultimately, the miners working under the mountain during the slide were unharmed and dug themselves out (Cruden & Lan, 2015): townspeople were the ones who died in this disaster. Based on contemporary and historical analysis, Cruden and Martin (2007) accordingly argue that the Frank Slide was not caused by any one determinant. Most scholarship devoted to it now points towards a convergence of unstable conditions: the mine and the town connected to it were simply in the way of terrible luck (Charrière et al., 2016; Haug, Rosenau, Leever, & Oncken, 2016; Humair, Pedrazzini, Epard, Froese, & Jaboyedoff, 2013). The latest geological research argues that the Slide was attributed to an assemblage of (a) geological conditions, (b) fluvial erosion, (c) hydrometerological and hydrogeological conditions, (d) weathering processes, (e) seismic loading, and, finally, (f) anthropic influence (Charrière et al., 2016).

Since the 1980s, better remote sensory technologies have been installed into pressure points on Turtle Mountain (Fraser & Gruendig, 1985). The Turtle Mountain Monitoring Report (Alberta Energy Regulator, 2016a) indicates that remote sensors were divided into primary, secondary, and tertiary sensors. Primary sensors worked year-round to monitor easily detectable movements and vibrations using various tiltmeters, extensometers, and crack gauges. Secondary sensors measured movements but took environmental factors into consideration, with the resulting data fed into the Frank Slide Interpretive Centre by GPS devices. Meanwhile, tertiary sensors could detect deformations beneath the surface of the mountain (unlike the other two) using a vibrating-wire piezometer and thermistors positioned to measure pore-pressure in the rocks to determine movement and vibrations beneath the surface (Alberta Energy Regulator, 2016b).

The publicly funded installation of more high-definition remote sensors has been crucial for the accurate prediction of another slide; indeed, high-definition remote sensors are a universal necessity for all disaster sites (Pesci et al., 2007; Tarchi, Casagli, Moretti, Leva, & Sieber, 2003). Remote sensors take what was once a site's potential for future disaster and accesses the possible, thereby identifying the approximate date and speed of such disasters and allowing time for a proper evacuation in the event of recurrence. What is most imperative for disaster sites (especially ones that are also cultural heritage sites) is simply to anticipate and to communicate such anticipation to those who will be affected.

Landslides are certainly the easiest of disaster sites to monitor, with mostly accurate predictions (Sassa & Canuti, 2008): the speed, efficiency, and real-time analysis of changes to the surface of landslide sites has greatly benefited communities, giving them warnings of events from six hours to several weeks prior to occurrence (Alexander, 1991). Of late, people can receive warnings in real time through information from data loggers that can be uploaded as alerts directly onto people's smartphones and into email accounts. This is certainly the most imperative task for the sensor innovation technology, as well as for geotechnical developments—that is, wireless systems that communicate effectively and quickly without running out of battery life (Nguyen, Thanh, Nguyen, & Huynh, 2015).

Representational Milieux

I turn now to explore the cultural themes that underlie a disaster/tourism site, since disaster sites tend to invite the "tourist gaze." As John Urry wrote, they are "structured by culturally specific notions of what is extraordinary and therefore worth viewing" (Urry & Larsen, 2011, p. 59). Cultural life is not an addendum to the science and history of the Frank Slide. Rather, it is significantly entangled within and at the edges of human activity; it embraces the contours of the scientific and the natural disaster, transforming the disaster into several overlapping intensities and temporalities, which become a point of entry for leisure and for meaning-making. It is noteworthy that landslides and natural disasters are often taken as interrupters of cultural heritage, and not necessarily sites of cultural heritage themselves (Bromhead, Canuti, & Ibsen, 2006; Canuti, Margottini, Fanti, & Bromhead, 2009). Disaster sites and the like are meaningless without the culture or politics invested in them (Guggenheim, 2014). In understanding this, sociologists have turned more recently to the tools, technologies, and conceptual apparatuses that aid understanding of the interwoven local, national, and international lives that bring meaning to such places.

Sciences and social science researchers increasingly investigate disaster sites, research that has meant death prevention has improved greatly due to diminished risk and severity at these sites (Guggenheim, 2014). However, the cultural aspects of disaster sites tend to fall within dark tourism (Bowman & Pezzullo, 2010): namely, their cultural allure lies partly in their safe proximity to experiences of death. John Urry and Jonas Larsen (2011) commented on this allure of death in certain tourist destinations:

> These places of death, disaster and suffering have come to be performed as places of leisure, often charging an entrance fee, providing interpretation and selling various other services and souvenirs. Many of these places developed and continue because of well-organised enthusiasts and fans [...] These enthusiasts perform "work" involving reciprocity and mutual aid [...] Organized fans or enthusiasts bring this experience of death and disaster into the public eye, to make the world witness it through a public commemoration. (p. 219)

The Frank Slide is itself a unique site that resists such a designation—even while it is a disaster site, it is also a heritage site and a municipality. Thus, the cultural perspective facilitates examination of characteristics of *vitality* in disaster sites, and of multiplicity in time, space, and life, beyond that which is theorized as "dark." Of course, this is not to deny that the Frank Slide is popular with tourists, who visit the interpretive centre regularly and stand by the highway turnouts to inhabit a broad range of perspectives and sensations, snapping selfies or gliding panoramas over the image-surface of bone-crushing destruction. This adds a dimension of pleasure to the site, much of which is engaged with destruction and awe: "The destruction on the valley floor is jaw-dropping," writes one *Yelp* reviewer. "Pity to the unfortunate souls who lived in the path of this event of gruesome raw power" (Corey, 2009, https://www.yelp.ca/biz/the-frank-slide-crowsnest-pass).

The poetic imagination also comes to represent the Slide. Examples include Robert Gard's *Ballad of the Frank Slide* (1949), famed singer Stompin' Tom Connors' boot-stomping *How the Mountain Came Down* (1969), and *Frank AB* by the Rural Alberta Advantage (2008/2009). The Frank Slide, as it is commemorated within such fictionalizations, stands as a prescription for living in that space, understanding it, a spatial code that is more than just reading or interpreting.

According to Katie Gough (2012), "In times of natural disaster we are not often asked to consider what embodied cultural memory means, or how and where it may exist," since we have less time to contemplate than to mourn over the "pictures akin to an apocalyptic crime scene alongside a body count" (p. 103). Gough writes that cultural memory and performance rituals are intended to circumvent the tendency for nature to appear in its bare "state of emergency"; the culture of disaster is then erected to prevent nature from entering into a "zone of exception" that Giorgio Agamben (2004) considered central to contemporary politics. Thus, songs such as those named above offer a poetic buffer that prevents us from accessing the Slide's terrifying reality. Anthropomorphizing ecological disaster places human will and reason as a strategy for anticipation, comparable to the remote sensors, and government and industry funding.

An architectural component also participates in constructing a heritage site out of a landslide, or rather *into* a landslide. The Frank Slide Interpretive Centre sits comfortably at the northwest curve of the fan, hosting tourists (but rarely local residents) to a range of learning activities related to all the scientific and cultural aspects of the Slide. The three floors are built to resemble collapsed shale and limestone, and contain technologies from seismic reading sensors to pull, push, and jump on, to puzzles and drawing rooms for children. Other disasters known to the Crowsnest Pass, such as the Hillcrest Mine explosion (that claimed the lives of 100 miners) are also commemorated. The last exhibit is a replica skull of "Black Beauty" (Currie, 1993), the black tyrannosaurus skull that was found some 15 km away in the Crowsnest Pass riverbend.

Snaking around the interpretive centre is a 10 km path through the rocks. It becomes clear on the walk that these are not "rocks" like one finds near a riverbed or in their back yard, but two-storey chunks of mountain face that a century ago loomed 3000 feet over the earth, never touched by anything but the elements for millions of years. Today, they are walked through, touched, and climbed on, as they are popular for bouldering, a sport that has become so sought-after in the Slide that local boulderers have held a "Tour de Frank" annual bouldering competition among the jagged rocks.

The area is also home to the pika, a small mountain mammal (about six inches in length) that only lives between boulders that are usually distributed across mountaintops. Recent research suggests that these creatures are reacting to climate change by moving towards cooler temperatures, an "upslope range retraction" (Moritz et al., 2008). But on the Frank Slide floor, the pikas are unaware that the top of their mountain is, for people, the base. By chance, they remain capable of caching food sources, which has a place in the Slide by virtue of having been placed there.

Similar to other cultural heritage sites built around natural disasters, like New Orleans' post-Hurricane Katrina bus tours or the city of Pompeii, this disaster interpretative centre represents the culmination of the interweaving of the past, the present, and the possible—co-constituted through science, industry, and culture. As such, disaster sites easily become overwhelmed with their own past; their temporal surplus spills into the future as a construction of the foreseeable. In this way, these sites become rationally explicable. While they may focus on (re)creating "emergence," their representational aspects or recreations have to do more with the recreation of the *event* than with the virtuality of the *emergence*. As Anders Ekström (2017) writes, disasters such as Pompeii are heavily mediated at their tourism sites with movies, twitter feeds, and other types of "remediation" that are intended to recreate the event. These modalities remind us that the re-temporalization of events depends on the continuum of the past, present, possible, or even inevitable, fanning out before our gaze.

Rhythm Revisited

Together, these disaster sites and their related interpretive events or facilities can be understood as milieux that constitute an infrastructure of audiblity for the imaginary organ of disaster, as it is played throughout the imaginary of a cultural heritage site. They may be placed exclusively in any one of these categories—they are rather concrete abstractions all at once, whose realities are lived out in place. The notion that a disaster such as the Frank Slide happened in the past and will happen again overlooks how such an event is also a unique production that unfolded atop a surface that is continuously changing. As mentioned above, the Slide is always moving, as geologists and residents in the area agree the mountain is not static. This instability is now monitored by remote sensors that extract movements imperceptible to the naked human ear or eye (though one can easily hear boulders and debris cascading down Turtle Mountain's empty face on a normal day), a need that highlights the severity of a disaster such as the Frank Slide. This threat is a continuing resonance, a form of "bare nature," a notion that I borrow from Rob Shields (2013) but with a slight variation: whereas Shields constructs the term to frame the manner in which resource extraction excludes nature from a semiotic framework, the sense of bare nature here is a suddenly sublime event involving a pure force of nature. As its own forceful event, it excludes itself from a semiotic reading. *Inclusion* in the semiotic system is beyond our capacity for understanding, since this would demand making *sense* of a paradox.

Technological objects vitalize the rhythm of place. For instance, the ground-based ISAR is a small, ample, and globally connected technology, with a geoengineering capacity that might spark the perennial fears of social scientists about the control of nature, environmental governance, and the "revenge of the Anthropocene" (Rickards, 2015). However, as Nigel Clark (2013) contends, a new "geologic politics" is one that urges social scientists to resist the enduring narrative of "defensive and reactive responses to geoengineering" (p. 2831). He suggests they vitalize ecology by learning "to think creatively and speculatively around interventions in earth systems—as well as engaging critically. They must be willing to reengage, in new ways, with a long human history of active, hands-on intervention in valued physical systems" (p. 2331). The lens of dark tourism might have the Frank Slide make a spectacle of tragedy. But such a framework is but one temporality of many that make up a cultural heritage disaster site.

As a postmodern subset of the tourist industry, dark tourism is often-times framed as a fetishization (both as commodity and as sexual) of the abject: it is assumed to have a negative and one-dimensional con-notation. The Frank Slide would certainly fit within the annals of dark tourism, which Anthony V. Seaton (1996) calls a thanatopic-based indus-try founded on the fascination with human- or naturally caused disas-ters and death. Other examples here include sites like Auschwitz, New Orleans, and 9/11 (see Potts, 2012). Obviously, from this perspective the Frank Slide is thus more than a "heritage site." By way of compar-ison, Paul Antick's (2013) fictionalization of the Bhopal disaster site's conversion into a tourist destination brings into focus just how unfore-seen disasters carry a tragedy-cum-profit potential that can contribute to a global and local economy. This designation exemplifies spectacular sites of disaster-commodification, such as 9/11. However, according to Tracey Potts (2012), this designation ignores the broader context, which must take into account "the circumstances of reception and consump-tion…to deal adequately with the matter of politics" (p. 245).

These "extreme objects" of dark tourism (Potts, 2012, p. 246) are related to Urry's discussion of "tourist kitsch" (Urry, 2003, p. 8), where disasters "are no longer there as such and can only be seen through their images" (p. 55). But in the case of the Frank Slide, which mixes tourism with educa-tion and scientific research, such a place could instead reshape encounters with death by associating those encounters with hope, possibility, and reinvention. Thus, while dark tourism assumes a negative and flatten-ing experience, the possibility of connecting death to resilience, memory, and community resonates strongly in the Frank Slide heritage site.

This particular cultural heritage site is not just an optically fascinating site, but takes other non-ocular senses and sensations into account, such as the force of the Slide's collapse, its thunderous sound, its duration, or its tactility. This foregrounding of non-ocularcentric representation— inviting visitors to play on seismic sensors, to climb on the rocks, and so on—reflects a "proximal construction of touch" that Kevin Hether-ington (2003, p. 1936) has proposed goes into spaces of representation and museums. It is a clear contrast to the typical distal knowledge (i.e. knowledge that is prescribed vs. internalized) often found in such sites. In proximal knowledge, the tactile and the haptic feed an embodied and performative knowledge production, giving rise to a more fluid form of knowing: "Proximal knowledge is embodied, sensory and *unsightly*—it implies an out-of-the-way, out-of-sight approach to knowing the world" (Hetherington, 2003, p. 1935).

Place is made, in part, through its ongoing fictionalization. In the case of the Frank Slide this includes these examples:

- *Film*—A film retells the event of the Slide and a slideshow presentation covers the history of Alberta coal-mining industry.
- *Music*—Several well-known Canadian songs go from the literal retelling of the Slide, as in Stompin' Tom Connors' song, to the metaphoric ballad of the Rural Alberta Advantage.
- *Literature*—This includes graphic novels and local works of fiction available at the Frank Slight Interpretive Centre.

Place is also part of the geological research, which uses digital technologies to recreate what the Slide might have looked like before the fall and then builds narratives from research and learning materials.

Revitalization makes the work of living imperceptible. Three modes of revitalization have so far been mentioned in relation to the Frank Slide site: human (boulderers), nonhuman (pikas and their habitat), and unhuman (rock and debris). In contrast to the representational function of fiction-alization, these direct relationships to the Slide are oriented towards the imperceptible, which manifests in multiple ways. The imperceptible is an act performed by human bodies that hug the contours of the boulders by climbing on paths otherwise concealed from easy viewing. Similarly, the pikas who invade the Slide are well fed on their natural food source, which elsewhere grows on mountaintops; here, the Slide performs this pseudo-mountaintop. For these living bodies, the rocks are habitat, at once alien but habituated to and inhabited.

Regarding the unhuman, the boulderers, for instance, immerse themselves in what Sally Ann Ness (2013) calls an "ecology of falling," less a consequence of climbing than the inevitability of it. Boulderers, for instance, generally anticipate falling, though they are not held close to the rock by a belay; they instead fall directly onto strategically placed foam pads while peers are poised, arms extended, ready to "take" the fall *away* from the climber. On the Frank Slide the prospect of falling is more ubiquitous, with boulderers warning that "Frank will *eat your stuff* if you drop anything" and that the "ground is not stable and *will* move around," warning repeatedly that "you are falling into a pile of jagged rocks; not soft sand" (Vallee, 2018). Here, we see a continuous process of "falling" less as a politics of failure and more as conjoining with the living logic of the disaster. And at the start of a recent *Tour de Frank*, after the boulderers waited at a meeting table for instructions for their competition, they dispersed amidst cheers, one laughing at the rocks twisting and creaking under her feet: "It's all going to fall out from under me" (Vallee, 2018). They disappeared, imperceptibly, into a winding crevasse of rock, into the silent trails concealed beneath the tourist-designated pathways.

A place is therefore not singly determined. In fact, here, the Frank Slide's defining marker is not the now-disproven bad mining practices deep within its recesses (indeed, the mine was one of the few parts of the emergency zone which withstood the collapse). Instead, place issues forth through a series of unforeseen and ongoing happenings. These can be considered "ontogenetic lines" that cut into one another, spring new assemblages, and attend more to those creatures who dwell in or use this geographic space than to the rapid and irrational speed at which the Slide exploded onto the town below. Place is thus an ongoing, epistemic, and diversified network of possibilities, embedded within a place's vitalization, devitalization, and revitalization. These possibilities are supported with technological innovations that enable movement within a place. In the case of the Frank Slide, the positioned telescopes on the path allow tourists to gaze on the peaks of Turtle Mountain.

The Imaginary Organ of an Impossible Time

It is a well-known argument that disaster sites establish a spurious dichotomy between the past and the present—a real past event is commodified, which risks the distortion of a past tragedy for a present gaze. Hurricane Katrina in New Orleans, 9/11 in New York City, the concentration camp at Auschwitz are obvious examples here. J. John Lennon and Malcolm Foley (2000) consider dark tourism a reflection of the postmodern condition in their foundational work on this notion, noting that sites intertwine the local with the global. Further, situated events, such as the failure of the Titanic, the tragedy of Auschwitz, or the losses of Frank Slide raise doubts about the triumphs of modernity. Meanwhile, they also incorporate pleasure and education in the commodity-object that results at these sites. In this way, then, dark tourist sites become places of lifestyle consumption.

Conversely, Maximiliano E. Korstanje (2016) has argued that dark tourism encompasses a heightened spirit of modernity in its emphatic fascination with death as a point of enjoyment for tourists: they learn through past tragedy that civilization, despite its struggles and tragedies, remains triumphant in its reinventions and especially its measures for the prevention of the reoccurrence of tragedy. Thus, this introduces the future as a preventable temporal dimension. Both positions are interesting because they produce unique specific orderings of time: the past either raises doubts about the abilities of civilization or past tragedy reinforces better future triumphs, replete with preventative techniques. Temporality is essentially objectified by the sites themselves, even as these disaster sites, *as* cultural sites, presume an overarching thematic narrative of civilization (whether as successful or otherwise).

Nonetheless, at the Frank Slide Interpretive Centre, the anticipation of future disaster is built into the narrative framework as much as is the past event. Indeed, ecological disaster is all around us, which positions the Frank Slide as a harbinger of a coming catastrophe that today is grounded in climate change. It becomes a past echo ringing into the future, where the sustainability of human life is under threat, a condition that lays to rest the debate about whether we are modern or postmodern—catastrophe cares little for such distinctions. However, the appeal of disasters may point more to people's fascination with an empirically founded finitude than to the nature of our accomplishments. Ekström (2017) argues, for instance, that the debate over the artificiality of past authenticity at disaster sites is diminishing as visitors accept that past images are meaningful mediations of past liveness. This is because at such sites the disaster is immediately resonant, felt as palpable.

To consider how a place moves on heterogeneous time scales (from the slow scalar movements of mountains to the scurry of a pika), we must reach for a dynamic conception of place as a site of ongoing negotiation and change, rather than as a place solely determined by a particular place-myth or type of memorialization. In disaster sites, the memorialization produces a dominant place-myth that engenders place as static: a constant place, where its spatial and temporal dimensions capitulate to a dominant narrative (in our case study, "Canada's deadliest rockslide"). This familiar myth is reinforced in much of the interpretive centre's architecture and display, built with personal and national memories of the mountain that would eventually become a slide, the foremost disaster of the Crowsnest Pass.

The spatiotemporal nature of a disaster event is the singular defining feature that aligns such sites with other sites of dark tourism: they are a pilgrimage into the spectator's distanced gaze. This is how Massey (2005) describes a spatiotemporal event at least, where the occurrence is specified by the stories that have been gathered around it and circulate within it; it is a multiplicity of social perspectives performed and maintained through multiple interactions. But Massey points to another contradistinctive category: *event of place*. This genus is defined less by the collusion of narratives that maintain a dominant spatiotemporal event, and more by "simply a coming together of trajectories." In the event of place, the site is open to constant variation and mutation, as well as actions involving human and nonhuman time scales and imperceptible vibrations and movements. Like the mountain that moves, movements are as gradual as the shifting continents, so Massey's event of place is defined by a *throwntogetherness* of incoherent and contingent trajectories.

Massey (2005) is not necessarily interested in the contingencies of place, as though all trajectories lead to one defining yet unforeseeable event. Rather, she is more interested in the ongoing production of place as it is made up by multiple "stories so far." These stories comprise all subjective statements on place, those that are closely related as much as those that are eternally disconnected. Thus, while a spatiotemporal event is contained within a sense of the immediate present and the humans who tell those stories, a more expanded event of place refers to those events that are well beyond human perception or sensation, though include it. These intersecting trajectories constitute this event of place. Here, it is bound by that expanded *throwntogetherness* that refers to all entities and matter and temporalities and contradictions. Different times (past/present/possibility/potential/finite/infinite), matter (solids/liquids/gases), space (Euclidean/non-Euclidean), social groupings, normative structures, hierarchies, and so on engage in an ongoing "negotiation which must take place within and between human and nonhuman" (2005, p. 140). Further, they are a "constellation of processes rather than a thing" (p. 140). The analyst is called to find how these components fit together in a coexistence of stasis and movement.

Massey (2005) argues also that temporality is a unique frame of reference for place, to the degree that time is measured as a pure relation between interacting social groups; this sociality creates a sense of place that is both local and global. Her sense of time makes it possible to theorize place through human and nonhuman interactions if those interactions, however heterogeneous, converge on a place. The differences between geological and human time cycles lead Massey to conceptualize place as "heterogeneous associations." For example, she conceptualizes "immigrant rocks" as an example of how the imperceptible tectonic movements deliver rocks as passersby witness a much faster world of human time. This example points to the impossibility of defining place according to any one feature, since rocks are transient visitors whose mobilities span millions of years. She writes about this in these terms:

> This is the event of place. It is not just that old industries will die, that new ones may take their place. Not just that the hill farmers round here may one day abandon their long struggle, nor that that lovely old greengrocer's is now all turned into a boutique selling tourist bric-à-brac. Nor, evidently, that my sister and I and a hundred other tourists soon must leave. It is also that the hills are rising, the landscape is being eroded and deposited;

the climate is shifting; the very rocks themselves continue to move on. The elements of this "place" will be, at different times and speeds, again dispersed. (Massey, 2005, pp. 140–141)

Given the rise in environmental uncertainty and the very real anticipations of ecological catastrophes under the conditions of climate change, Timothy Morton (2013) has recently argued that we need to rethink imperceptible mobilities (such as tectonic shifts) as *hyperobjects*—directly felt local consequences of broader imperceptible cyclical change. Hyperobjects, directly encountered and therefore incapable of being metaphysical, paradoxically appear larger with the greater objective distance one has from them (the grey smudge of Frank Slide on Google Earth, for instance). This is what makes the hyperojbect a finite force, rather than an infinite one. This *very large finitude* can generate a new type of anxiety related to anticipation of certain, sweeping change (like that specific to environmental uncertainty). Morton writes of this:

I can think infinity. But I can't count up to one hundred thousand. [...] It's unimaginably vast. Yet there it is, staring me in the face, as the hyperobject global warming. And I helped cause it. I am directly responsible for beings that far into the future, insofar as two things will be true simultaneously: no one then will meaningfully be related to me; and my smallest action now will affect that time in profound ways. [...] There is a real sense in which it is far easier to conceive of "forever" than very large finitude. Forever makes you feel important. One hundred thousand years makes you wonder whether you can imagine one hundred thousand anything. (p. 71)

Morton (2013) writes that we are only capable of encountering the very large finitude because of our inheritance of an Einsteinian space-time theory:

Einstein's discovery of space-time was the discovery of a hyperobject—the way in which mass as such grips space, distorts it from within, stretching space and time into whorls and vortices. For Einstein, entities—which may or may not include living observers—comprise indivisible "world tubes." By *world tube*, relativity theory means to include the apprehending aspect for an entity with its entitive aspect. A world tube is a hyperobject. That is, world tubes stretch and snap our ideas of what an object is in the first place. Each world tube encounters a fundamentally different universe depending on its mass and velocity. World tubes recede from other world tubes in an inescapable and irreducible way. (2013, p. 56)

Morton approaches time and space as productions of real physical events: multiple times and multiple spaces, unbound by a more hegemonic commodity-time or time of dark tourism. Nothing is faster than the light that brings to perception the information that the object contains its situated geographical and historical information, which is its space and time. For instance, a rubble of rock emits a temporality that produces an unbounded space and time, even as it contains the very brute fact that mountains move over millions of years on the earth, and pass from one perspective to the other and can inhabit multiple timescales, even while we are ensnared by the vortices that this multiple perception affords us, since each timescale has its own gravitational pull. With the Frank Slide, the geographical and historical slower timescales and space-time are made manifest in the acoustic fluidization that produced the explosive and unique fan-like appearance of the Frank Slide.

This perspective of the very large finitude demonstrates a new capacity to think through places as finite but still grapple with their geohistorical emergences. An abyss lies before hyperobjects, but it is not an infinite abyss—it is one produced through geohistorical manifestations. This is why the multiple timescales produced in the Frank Slide Interpretive Centre represent both one timescale and more than the one. Morton (2013) argues that objects produce their own conditions of unthinkability, the "monstrous" so to speak, the compelling horror, akin to the dark tourism of the Frank Slide, which is what summons humans and nonhumans alike into the dimension of imperceptible movement. In contrast to an object's permanence, Morton's approach to time and place proposes that new media technologies allow us to simultaneously experience multiple time dimensions, to exist within the bends and curvatures of time that ripple across the surface of things. He writes: "Once we become aware of it, undulating temporality corrodes the supposed fixity of smaller objects that lie around me."

As with Massey's (2005) event of place, the hyperobject perspective makes redundant the idea that space is a container for things and that time is an empty vessel from one event to the next spatiotemporal event. It is perhaps because the very large finite produces the most unthinkable aspect—that is, that the very large finite will be here long after the human race is extinct—that the Frank Slide will bear no witness. Not even the mountains can do this since their destination will be eventually in the same pile of rubble, sometime between now and thousands of years from now. Even designating a site as rubble is a human evaluation, but the mountain, bifurcated into both rock and slide, possesses a time witnessed by the rocks. This is what Morton (2013) calls the "future future":

> If time is not a neutral container in which objects float, but is instead an emission of objects themselves, it is at least theoretically more plausible that an object could exert a backward causality on other entities, than if objects inhabit a time container that slopes in one particular direction. (p. 67)

Concluding Remark: Time in Place

Part of my intent with this chapter has been to demonstrate, by way of a brief and partial sketch of the Frank Slide, how place conjoins multiple temporalities and time horizons into a rhythm of place, like the stretch of time representations for science and the infrastructural time practice of human/nonhuman encounters. I did not attempt here to construct any new definitions of place, nor claim that place is now dominated by a new ideological narrative. However, it is increasingly clear that places are never singular, but are ever multiple. As such, they require a resonant sensibility so as to locate their dynamic and continuous variation, including their deformations and their mutations. Such a perspective resonates markedly with the cultural topologies appearing in the annals of cultural theory (Lury, Parisi, & Terranova, 2012), which emphasize the notion that the essential features of space persist, despite the continuous mutations of the elements that demarcate its boundaries.

As I have argued above however, every place is unique, made of smaller other places—a place for boulderers is influenced by the Frank Slide, for example, even while boulderers are only marginally interested in the Slide "itself." If the place-myth of the Slide is sublimated into the nation by its place-image (here, as noted "Canada's deadliest rockslide"), it stands to reason that resistances to this exist. Instead, we have a continuous variation on a theme, that of *vitality*. But something more temporal is involved in place-making, including memorial activities and antimemorial, present-focused activities such as bouldering. As well, some temporally oriented activities encompass only the future of the present moment as a desired memory (such as a selfie by the rocks). The visitors, the tourists, the boulderers, and of course the researchers and the interpreters all maintain a relationship of finitude to such a place of crisis awe, an epiphany of catastrophe. There is no denial of the future, nor any denial of history; rather, overlapping temporalities maintain powerful relations to place.

References

Agamben, G. (2004). *The open: Man and animal.* Palo Alto, CA: Stanford University Press.

Alberta Culture and Tourism. (2016). *Frank Slide Interpretive Centre: FAQ—Frequently asked questions.* Retrieved from http://www.history.alberta.ca/frankslide/faq/faq.aspx.

Alberta Energy Regulator. (2016a). *Turtle Mountain monitoring program.* Retrieved from http://ags.aer.ca/turtle-mountain-monitoring-program.htm.

Alberta Energy Regulator. (2016b). *Historical monitoring at Turtle Mountain.* Retrieved from http://ags.aer.ca/historical-monitoring-at-turtle-mountain.

Alberta Energy Regulator. (2016c). *Alberta geological survey: Current monitoring.* Retrieved from http://ags.aer.ca/current-monitoring.

Alexander, D. (1991). Information technology in real-time for monitoring and managing natural disasters. *Progress in Physical Geography, 15,* 238–260.

Ang, I. (2006). From cultural studies to cultural research. *Cultural Studies Review, 12,* 183–197.

Antick, P. (2013). Bhopal to Bridgehampton: Schema for a disaster tourism event. *Journal of Visual Culture, 12,* 164–185.

Benko, B., & Stead, D. (1998). The frank slide: A reexamination of the failure mechanism. *Canadian Geotechnical Journal, 35,* 299–311.

Bode, R. (2014). Rhythm and its importance for education. *Body & Society, 20*(3–4), 51–74.

Bonnard, C. (2011). Technical and human aspects of historic rockslide-dammed lakes and landslide dam breaches. In S. G. Evans, R. L. Hermanns, A. Strom, & G. Scarascia-Mugnozza (Eds.), *Natural and artificial rockslide dams* (pp. 101–122). Berlin, Germany: Springer.

Bowman, M. S., & Pezzullo, P. C. (2010). What's so "dark" about "dark tourism"? Death, tours, and performance. *Tourist Studies, 9,* 187–202.

Bromhead, E. N., Canuti, P., & Ibsen, M.-L. (2006). Landslides and cultural heritage. *Landslides, 3,* 273–274.

Canuti, P., Margottini, C., Fanti, R., & Bromhead, E. N. (2009). Cultural heritage and landslides: Research for risk prevention and conservation. In K. Sassa & P. Canuti (Eds.), *Landslides: Disaster risk reduction* (pp. 401–433). Berlin, Germany: Springer.

Charrière, M., Humair, F., Froese, C., Jaboyedoff, M., Pedrazzini, A., & Longchamp, C. (2016). From the source area to the deposit: Collapse, fragmentation, and propagation of the Frank Slide. *Geological Society of America Bulletin, 128,* 332–351.

Clark, N. (2013). Geoengineering and geologic politics. *Environment and Planning a, 45,* 2825–2832.

Corey, G. (2009). *Now this is a site to see.* Retrieved from https://www.yelp.ca/biz/the-frank-slide-crowsnest-pass.

Cruden, D., & Lan, H. X. (2015). Using the working classification of landslides to assess the danger from a natural slope. In G. Lollino, A. Manconi, J. Clague, W. Shan, & M. Chiarle (Eds.), *Engineering geology for society and territory* (Vol. 2, pp. 3–12). Basel, Switzerland: Springer.

Cruden, D. M., & Martin, C. D. (2007). Before the Frank Slide. *Canadian Geotechnical Journal, 44,* 765–780.

Currie, P. J. (1993). Black beauty. *Dino Frontline, 4,* 22–36.

Degnen, C. (2005). Relationality, place, and absence: A three-dimensional perspective on social memory. *Sociological Review, 53,* 729–744.

Deleuze, G. (2003). *Francis Bacon: The logic of sensation.* London: Continuum.

Deleuze, G., & Guattari, F. (1987). *A thousand plateaus: Capitalism and schizophrenia* (B. Massumi, Trans.). Minneapolis: University of Minnesota Press.

DeLyser, D., & Sui, D. (2013). Crossing the qualitative-quantitative divide II: Inventive approaches to big data, mobile methods, and rhythmanalysis. *Progress in Human Geography, 37*(2), 293–305.

Ekström, A. (2017). Remediation, time and disaster. *Theory, Culture & Society, 33*(5), 117–138.

Fraser, C. S., & Gruendig, L. (1985). The analysis of photogrammetric deformation measurements on Turtle Mountain. *Photogrammetric Engineering and Remote Sensing, 51*, 207–216.

Gardiner, M. E. (2012). Henri Lefebvre and the 'sociology of boredom'. *Theory, Culture & Society, 29*(2), 37–62.

Gough, K. M. (2012). Natural disaster, cultural memory: Montserrat adrift in the black and green Atlantic. In W. Arons & T. J. May (Eds.), *Readings in performance and ecology* (pp. 101–112). New York, NY: Palgrave Macmillan.

Government of Alberta, Alberta Emergency Management Agency. (2017). *Emergency response protocol for Turtle Mountain.* Retrieved from http://www.aema.alberta.ca/documents/Emergency-Response-Protocol-for-Turtle-Mountain.pdf.

Grosz, E. (2008). *Chaos, territory, art: Deleuze and the framing of the earth.* New York: Columbia University Press.

Guggenheim, M. (2014). Introduction: Disasters as politics–politics as disasters. *Sociological Review, 62*(Suppl. 1), 1–16.

Hall, T., Lashua, B., & Coffey, A. (2008). Sound and the everyday in qualitative research. *Qualitative Inquiry, 14*(6), 1019–1040.

Harrison, T. W., & Friesen, J. W. (2015). *Canadian society in the twenty-first century: An historical sociological approach* (3rd ed.). Toronto, ON, Canada: Canadian Scholars' Press.

Haug, Ø. T., Rosenau, M., Leever, K., & Oncken, O. (2016). On the energy budgets of fragmenting rockfalls and rockslides: Insights from experiments. *Journal of Geophysical Research: Earth Surface, 121*, 1310–1327.

Henriques, J. (2014). Rhythmic bodies: Amplification, inflection and transduction in the dance performance techniques of the "Bashment gal". *Body & Society, 20*(3–4), 79–112.

Hernández-Moreno, G., & Alcántara-Ayala, I. (2016). Landslide risk perception in Mexico: A research gate into public awareness and knowledge. *Landslides, 14*(1). Retrieved from https://doi.org/10.1007/s10346-016-0683-9.

Hetherington, K. (2003). Spatial textures: Place, touch, and praesentia. *Environment and Planning a, 35*(11), 1933–1944.

Humair, F., Pedrazzini, A., Epard, J. L., Froese, C. R., & Jaboyedoff, M. (2013). Structural characterization of Turtle Mountain anticline (Alberta, Canada) and impact on rock slope failure. *Tectonophysics, 605,* 133–148.

Korstanje, M. E. (2016). *The rise of thana-capitalism and tourism.* London: Routledge.

Lefebvre, H. (2004). *Rhythmanalysis: Space, time and everyday life.* London: A&C Black.

Lennon, J. J., & Foley, M. (2000). *Dark tourism.* London: Cengage Learning EMEA.

Lury, C., Parisi, L., & Terranova, T. (2012). Introduction: The becoming topological of culture. *Theory, Culture & Society, 29*(4–5), 3–35.

Marty, S. (2011). *Leaning on the wind: Under the spell of the Great Chinook.* Surrey, BC, Canada: Heritage.

Massey, D. (2005). *For space.* London: Sage.

McConnell, R. G., & Brock, R. W. (1904). *Report on the great landslide at Frank, Alta. 1903: Extract from part VIII, annual report, 1903.* Ottawa, ON, Canada: Government Printing Bureau.

Melosh, H. J. (1986). The physics of very large landslides. *Acta Mechanica, 64*(1–2), 89–99.

Moritz, C., Patton, J. L., Conroy, C. J., Parra, J. L., White, G. C., & Beissinger, S. R. (2008). Impact of a century of climate change on small-mammal communities in Yosemite National Park, USA. *Science, 322,* 261–264.

Morton, T. (2013). *Hyperobjects: Philosophy and ecology after the end of the world.* Minneapolis: University of Minnesota Press.

Ness, S. A. (2013). Ecologies of falling in Yosemite National Park. *Performance Research, 18*(4), 14–21.

Nguyen, T. D., Thanh, T. T., Nguyen, L. L., & Huynh, H. T. (2015). On the design of energy efficient environment monitoring station and data collection network based on ubiquitous wireless sensor networks. In *Computing & communication technologies—Research, Innovation, and Vision for the Future (RIVF), 2015 IEEE RIVF international conference* (pp. 163–168). New York, NY: IEEE Communications Society.

Pesci, A., Fabris, M., Conforti, D., Loddo, F., Baldi, P., & Anzidei, M. (2007). Integration of ground-based laser scanner and aerial digital photogrammetry for topographic modelling of Vesuvio volcano. *Journal of Volcanology and Geothermal Research, 162,* 123–138.

Potts, T. J. (2012). "Dark tourism" and the "kitschification" of 9/11. *Tourist Studies, 12,* 232–249.

Prior, N. (2011). Speed, rhythm, and time-space: Museums and cities. *Space and Culture, 14*(2), 197–213.

Rickards, L. A. (2015). Metaphor and the anthropocene: Presenting humans as a geological force. *Geographical Research, 5,* 280–287.

Sassa, K. (2016). Implementation of the ISDR-ICL Sendai partnerships 2015_2025 for global promotion of understanding and reducing landslide disaster risk. *Landslides, 13,* 211–214.

Sassa, K., & Canuti, P. (Eds.). (2008). *Landslides-disaster risk reduction.* Berlin, Germany: Springer Science & Business Media.

Seaton, A. V. (1996). Guided by the dark: From Thanatopsis to Thanatourism. *International Journal of Heritage Studies, 2,* 234–244.

Sharpe, C. F. S. (1938). *Landslides and related phenomena: A study of mass-movements of soil and rock.* New York, NY: Pageant Books.

Shields, R. (1991). *Places on the margin: Alternative geographies of modernity.* New York, NY: Routledge.

Shields, R. (2006). Flânerie for cyborgs. *Theory, Culture & Society, 23*(7–8), 209–220.

Shields, R. (2013). *Spatial questions: Cultural topologies and social spatialisation.* London, UK: Sage.

Simpson, P. (2012). Apprehending everyday rhythms: Rhythmanalysis, time-lapse photography, and the space-times of street performance. *Cultural Geographies, 19*(4), 423–445.

Stähli, M., Sättele, M., Huggel, C., McArdell, B. W., Lehmann, P., Van Herwijnen, A., … Springman, S. M. (2015). Monitoring and prediction in early warning systems for rapid mass movements. *Natural Hazards and Earth System Sciences, 15,* 905–917.

Stahr, A., & Langenscheidt, E. (2014). *Landforms of high mountains.* Berlin, Germany: Springer.

Tarchi, D., Casagli, N., Moretti, S., Leva, D., & Sieber, A. J. (2003). Monitoring landslide displacements by using ground-based synthetic aperture radar interferometry: Application to the Ruinon landslide in the Italian Alps. *Journal of Geophysical Research: Solid Earth, 108*(B8). Retrieved from https://doi.org/10.1029/2002JB002204.

Terzaghi, K. (1950). Mechanism of landslide. In S. Paige (Ed.), *Application of geology to engineering practice* (pp. 82–123). New York, NY: Geological Society of America.

Terzaghi, K. (1961). Stability of steep slopes on hard unweathered rock. *Geotechnique, 12,* 251–270.

Urry, J. (2003). *Global complexity.* Cambridge, UK: Polity.

Urry, J., & Larsen, J. (2011). *The tourist gaze 3.0*. London, UK: Sage.

Vallee, M. (2018, September 17). *Tour de Frank*. Field notes at the Frank Slide site.

Wachinger, G., Renn, O., Begg, C., & Kuhlicke, C. (2013). The risk perception paradox—Implications for governance and communication of natural hazards. *Risk Analysis, 33*, 1049–1065.

Wilson, D. (2005). *Triumph and tragedy in the Crowsnest Pass*. Surrey, UK: Heritage House.

7

Conclusion: Sounding Worlds

The book opened with the question, how does a world sound? The purpose of this conclusion is to ask a different sort of question: How does sounding world? Behind this question is a need to reconceptualize the ethics of bodies and their boundaries, their places and their trans-species communities. The book charted a trajectory from the voice of the individual as an imaginary organ discovered through the infrastructure of audibility in early laryngoscopy, to the proliferation of the voicescape across species and through technologies, the human/animal voice, the becoming computational of sound, and the rhythm of place. In all, sound and its variants were involved in the generation of new ethical relations within and beyond the individual.

Beginning with a sociotechnical imagination of the voice and the body, told through the development of the laryngoscope, and moving through the body into social and ethical trans-species relations and the rhythm of place, this book offered a series of encounters in service of the argument that sounding breaches the body's borders while opening onto a field of vibration. Sounding reaches into transformation. The project responded to the need for a theoretical repertoire that traverses those boundaries between

© The Author(s) 2020
M. Vallee, *Sounding Bodies Sounding Worlds*, Palgrave Studies in Sound,
https://doi.org/10.1007/978-981-32-9327-4_7

voice, sound, rhythm, and place: it captures the oscillation, the relations, and the vibrational, the refractions between science, subjectivity, objectivity, technics, and the places that those interactions and intra-actions produce. Such a language makes up for a general field of vibration, in which intimate entanglements in science, culture, and technology demonstrate inseparable components. The book thus carves a path towards the vibrational affects between organs, organizations, and organologies, to better access the multiple relations that bind the body's presence as an imaginary organ. I would like to end with a few words that trace the trajectory of the book from the point of view of voicing, sounding, and worlding, and again articulate a position for sounding that will open opportunities for future research in media theory, and science and technology studies.

Voicing

Voicing generates relations between bodies. Voicing is not the same as a voice. And, unlike voicing a concern, a complaint, or grievance (which stand more as examples of "giving voice" in the political sense), voicing is a way of addressing how technical intricacies condition the possibility for certain experiences, while simultaneously accounting for the spaces of indeterminacy in experiments, subject to variation and new directions. In a similar vein, the concept of voicing has less to do with contact or touching than with describing the conditions that make touching a possibility— voicing is the communicative possibility for touch. Much of the myth that we have inherited about the voice—that it is estranged from the body, for instance—has to do with our inherited impression that the presence of the voice is marred by its absence, as described in Chapter 3.

As a theoretical resource, the concept of a body-technic could potentially provide a means for understanding voicing from a media theory perspective, since a body-technic extends our understanding of embodiment beyond the enclosed and contoured human body. The question permeating this book pivots on ambivalences and ambiguities of mediation and the slipperiness of those mediations when they are voicings. The technologies through which the body creates worlds belongs to this notion of touching and contiguity. A body whose borders are expanding and collapsing in a continual oscillation is a porous and open body. The expansion and collapse of borders between bodies, things, ideas, and other beings (human, nonhuman, inhuman, more-than-human, less-than-human, unhuman) all belong to a burgeoning sense of what embodiment embodies. And what brushes the delineations of those boundaries before the rest of the body can catch up is all a matter of voicing. It is voicing which returns the body to itself but as something perhaps unrecognizable, again discussed in Chapter 3; it is like hearing one's own voice on a recording.

The body as body-technic is convincing in that it embraces the technological and seeks space beyond the body; further, it is convincing if it describes, not the bodily experience itself, but the condition in which certain experiences of the body are possible, which includes those in non-human encounters. It is also meaningful by being aligned with the posthuman project that understands the human as part of broader nonhuman configurations and agencies. Certainly, the body-technic maintains a focus on how the body's experience is preconditioned by the technics—which, in turn, open onto the uniqueness of other worlds.

However, even with the body's capacity to decentre human consciousness, the organs of the body (along with technical arrangements and organizations that produce those organs) still come to represent or substitute for those subject-centred experiences. The body-technic brings the body, its organs, its social organizations, and its technical frameworks into a dynamic generative structure that configures the body through its ongoing relations with the world. Organs, in turn, make it possible for the body to experience itself or the world. And through organs/extensions (like microphones, laryngoscopes, and so on) we get a sense of how we can broaden our definitions of "giving voice" to the body, so that they are not limited to the laryngealcentric, an earlier focus in this book that I migrated away from.

Voicing, within the framework of the body-technic, is not unavoidably a representation of the body—that is, a placeholder for the body's own projected representation. Nevertheless, it is true that the voice is one means by which a body reaches beyond itself. In sociological literature at least, we consider a marginal social group to have entered into representation when they "give voice" to a collective subjectivity. But here we are interested in the body-technic that breaks, reaches, breaches, experiments,

and touches all different matters and materials. Voicing is an embodiment that touches, while sounding is necessarily bound to infrastructural
embodiments. Touch is the transmission of energy that forces one surface
to proximate and change another. But touching can also imply surfaces
inside and beyond multiple surfaces, implied in an interior touching (the
feeling we may have as sound touches our hearing apparatuses or moves
through our bodies), or a touching to move. Where there is sounding,
there is touching, because sounding involves the vibration of a body as it
enters into the inhabited space of another body (or set of bodies). Thus, the
conclusion that the voice is an imaginary organ stands, in that it touches
without touching.

As discussed in Chapter 2, the human voice consists of a complex of
organs running from the lungs up through the head (indeed, one's gait
can be thought of as one of the organs involved in producing the human
voice). These organs produce an internal resonance that an individual has
an experience of, from the inside. If they are working as they should, they
produce a clear and distinct voice, the body breaking acoustically through
its own shell. However, the voice is not only a product of the organs,
whose specific hums, growls, pinches, and leaks escape into and through
every voice. When those organs permeate the boundaries of bodies, it
is unmistakably the voice that also articulates in sneezes, coughs, and
other involuntary gestures. But the voice also requires techniques that are
specific to its occupation of places: The voice of jokes whispered in church
is drastically different from the voice that calls a dog across a field to come
back to its master. Thus, voicing requires a collection of techniques for
embodying place, as well as for articulating those situated embodiments
through the vibratory complex of organs. Voicing requires not only a
chamber but a place in which the chamber resonates outward. Voicing is
thus a means of articulating the changes that occur between bodies and
the changes that come to a place as a result of voice.

Place, then, is another factor that should be included in our attempt
to come to terms with voicing. Place, as a singular event that is made
through the defined boundaries of space and time, enables voicing to
demarcate those very boundaries. My consideration of place throughout,

but especially in Chapter 6, makes it impossible to think of voicing as an intersubjective phenomenon; rather, it is more of a trans-subjective one. Place, to remain in place, requires techniques. Place does not give the voice the opportunity for voicing; instead, voicing places itself as a boundary that conforms with or goes against the techniques that make a specific place. Place is thus a mediational site, a singularity through which voicing connects bodies to one another. The technical is the group of practices that sustain and/or demand its transformation, as explicated in the becoming-animal of incorporeal transformation proposed by Deleuze and Guattari (1987) and discussed in Chapter 3.

At last, we arrive at the final formation that is produced through voicing, which is that of embodiment. In considering embodiment, we must be cautious, aiming to avoid a phenomenological description, and instead approach embodiment as a zone with expanding and collapsing boundaries. This is crucial to the concept of sounding. The purely phenomenological use of embodiment as the condition upon which experience is possible will not suffice here, nor will later adaptations that incorporated marginalized bodies into the purview of social change.

Embodiment here is a mediational point of transduction between the inner resonances of the body and those resonances' extensions into space and time that make up the singularity of place. Embodiment is thus understood as situated and situating: in situ. Embodiment resonates outward and inward, marking a place's situatedness. And voicing is no more or less than the total entanglements that reach across and between bodies, making up what I called in Chapter 4 the transacoustic community.

Voicing, as a singular concept, does not work. Instead, through the course of this book, I have explored a particular type of voicing in the voicescape and the transacoustic community. The voicescape is thus placed before us as a way that bodies transform themselves and others, and contribute to the making of place. The social, cultural, and political usages of voicescapes excavated in this book, especially in relation to embodiment and place, will be useful to those concerned with broader social issues. In particular, the transacoustic community is a new way of giving voice to individuals and groups across temporal and spatial boundaries. It presents a more generalized way of understanding how bodies are configured beyond themselves, as images held through voicings. The empirical data of this book are therefore far-reaching and widely instrumental, intended to push sounding and voicing forwards into new formations for understanding the body's technical complexities.

Sounding

Sounding is concrete, in the sense that it has come to refer to all of its participations: Sounding is as much in the mist-net as the microphone, as the landslide, as the boulderer's body. It is also an invitation to participate. Sounding opens speculation and wonder.

In this book, I have used sounding as a method for opening onto new ethical relations between human, technological, and other biotic and abiotic assemblages. Sounding subsumes the extraction and capture of vibrations so as to make sense of them—it refers to the labour that goes before, into, and above the capture of sound, into the kinds of orientations that sound opens up. Sounding refers to sound on the periphery of experience, while being nonetheless directly implicated in experience. Compatible with recent directions in sound studies, I have argued in this book that it is in our best interest to continue giving an ear to imperceptible vibrations, instead of sequestering sound to only that which is within the limited range of so-called natural human hearing. Imperceptible vibrations constitute a movement. All movement, perceptible and imperceptible, moves through an environment upon which its movement relies. To reiterate, the purpose here was to be sounding new openings through organisms and entities, with an ear to making imperceptible vibrations resonate with a robust undertaking of sound-based research in science, technology, and mediation.

Sounding, to be clear, is not the same as sonification, which refers to contemporary practices that organize data into sound (a transduction). For instance, John Luther Adams sonifies the collective sound of Alaska's geophysical movements in *The Place Where You Go to Listen* (see Ross, 2012), at the University of Alaska in Fairbanks. He uses synthesizers and lights connected to seismic readers across the state to perform an immersive, ongoing, and live composition whose composer is a vast, expansive, and complex landscape ecology. His installation resonates with and is composed alongside the dynamic movements of the earth, largely anchored in a specific place that provides the raw material. His music is specifically tied to the place it is in: "My music is going inexorably from being about place to becoming place" (quoted in Ross, 2012, p. 8).

But sounding, as it is proposed here, is grounded in the non-audible forces of place, the vibrations that bind a community, the place that resonates in the echo of a disaster, such as the very large finitude of the Frank Slide Heritage Site examined in Chapter 6. As I demonstrated in Chapter 4, the microphone is also one of the most important data collecting instruments for the sciences, explored in this book in concert with the becoming computational of sound. But unlike sonification, which Alexandra Supper (2014) describes in her analysis of John Luther Adams' music as "an auditory equivalent of data visualization in which data are turned into sounds" (p. 34), the computational paradigm of sound involves a radical non-listening that is less to be feared. Nor is it exemplary of Jennifer Gabrys's (2016a, 2016b) warning of the homogenization of the perceptible. Instead, as demonstrated in Chapters 5 and 6, the computational paradigm of sound contributes to the non-cochlear-centric and decentred definition of sound that retains undulating vibrations of experience outside of the human body. Those imperceptible vibrations directly correlate with the experience of the human body in its collective global habitat—indeed, increased biodiversity loss is said to have devastating effects on the human population.

Similarly, sounding is aligned with the recent turn in sound studies towards imperceptible vibrations, and thus is not particularly oriented to the physiology of hearing, or "phonocentrism" (Friedner & Helmreich, 2012). Certainly, the notion of sounding will not hold if we only approach sound as limited to human hearing (which picks up 20 Hz–20 kHz, below which are infrasonic vibrations and above which are ultrasonic vibrations). The purpose of sounding is therefore not to pursue sound as an object of knowledge, but instead as a more inclusive strategy, a broader take that aligns with acquiring a set of "sonic skills" (Bijsterveld, 2019)—which does not then reduce sound to the rather exclusionary "hearing" per se. The collective research into sound confirms that it is indeed pursued with an inclusive ethos, with researchers interested in transducing information for the benefit of widening perceptions. For instance, there is Joeri Bruyn-inckx's (2018) history of ornithology, and also Holger Schulze's (2018) anthropology of sound. What they share in common is an antireductionist approach to sound and to listening.

To approach the practice of sounding is to question the limits of visibility while accounting for the embodied practices of individuals who collectively work towards sounding. Sounding is decidedly practical. Two perspectives help in this matter of differentiating, yet complementary, definitions. Martin Heidegger's definition of sounding is that which hovers at the limits of representation, constituting "the character of strife (earth—world)" (quoted in Smith, 2012, p. 87). That is, his definition is meant to capture the stress between revealing and concealing (an object's sound gives it a contour, shape, and extension, all the while remaining enchained to the evanescence of sound). More recently, Julian Henriques (2008) has described sounding as a "kinetic activity, a social and cultural practice, a making and becoming" (p. 219). For our purposes here, these two definitions are entirely compatible. The practices that Henriques describes are useful for approaching the collective forces necessary for extracting imperceptible vibrations. The central issue around sounding is thus methodological.

How, then, do we think through the sound in sounding? Obviously, there must be some correlation between the two, for either to gain any measure of value from the other. Although it is tempting to suggest that sounding has always relied on an onto-kinetic frame, à la Heidegger and Henriques, I suggest that sounding holds a unique relationship with emerging sound technologies, which can capture a higher-resolution large data set of imperceptible vibrations through a coded and technologically bound interface. This capacity enables contemporary scientific research to use sounding as a method for detecting changes in biodiversity and the environment. In sounding, animals are defined by their acoustic properties over and above their visual phenotypes.

As demonstrated in Chapter 5, sounding thus propagates the following paradox: emerging sound technologies are capable of high definition and increased rate analysis, and require less human interference in the collection and analysis of data. Thus, a non-anthropocentric condition is found in sounding, in that actual sound waves matter; still, a renewed anthropocentric condition is also present because the purpose of sounding research is to reduce biodiversity loss. All this research depends on machines that listen to and analyze data. And the purpose of much of this research is to become *increasingly reliant* on technologies that listen in place of the human ear, since they hear what ears cannot. The movement towards autonomization is especially important to non-anthropocentric research, which decentres the exceptional listener and transforms the data into something programmable and autonomous. Sounding, then, by privileging this computational sound, disposes of the sound that can be heard by human ears.

Throughout this book, I have worked through the concept of sounding by aligning sound with the ecologies that attend to and generate sound data, bolstering the concept itself— constructing both sounding and sound as tensions at the horizon between world and earth, as Heidegger described sounding. With this, I argue that the "becoming computational of sound" leads us towards data analysis and the measurement of global populations in real time, which could generate a more global and trans-species connection between sound and place.

Towards an Elemental Sounding

Do we need sound for sounding? Or does sounding represent something more general underlying our interests in sound? How would we construct a more robust theory for sound and vibration, movement and embodiment, place and time, that diminishes its destination towards a listening event? This question hangs at the end of this book but is one that could be answered with new theorizations about the "environmental agencies" that produce subjectivities across biotic and abiotic systems. One of the most engaging readers on this topic has been Mark Hansen (2015), who has been particularly attentive to the broader implications of such a shift, arguing that media theory is in the midst of a migration from object-centred media studies towards the propagation of environments and networks. By way of this migration on the road to media environments he resolves that agency does not lie within the domain of any one autonomous social actor. In the spirit of other recent attempts at a posthuman media theory,

Hansen is cautious about the all-too-human ways that we theorize media and mediation, claiming that such a tradition is not compatible with the ubiquitous media of the twenty-first century. His perspective interrogates the elemental aspects of subjectivity, which calls out to its embeddedness—its "enworlding"—as an element among other elements that are of comparable value to the general agency and mutation of nodes in media environments; an easy example here are ubiquitous listening devices or voice assistants, such as Apple's Siri or Google's Alexa.

Hansen's (2015) interest is in preserving the concept of subjectivity, but not limiting it to human actors. We cannot talk about the decentralization of media and sensation without talking about the reconceptualization of subjectivity; thus, the interest in this book, particularly Chapters 6 and this chapter, in sound without listening, and the informationalization of sense. In environmentally based media theory, subjectivity might not be "subject-centred," as agency is distributed across networks but is not solely within the grasp of the subject. Freed of the contemporary compulsion to decentre or discard the notion of the subject, Hansen's preservational approach respects that contemporary subjectivity is formed and negotiated within the peripheries of direct experience. Subjectivity is therefore still wrapped into events as they unfold, despite the likelihood that those events are spatially or temporally distant from us. And these events are not representation of pre-existing realities—they are new sociotechnical imaginaries straddling data, culture, sense, and sound.

Thus, Hansen (2015) claims that elemental media, like sensors, produce a practical *doubleness*. They producing new worlds of sense that generate data, such as the backpack mic which transmits the inner vibrations of a nighthawk in flight (see Chapter 4), while also simultaneously producing new worlds *about* the data collected (i.e. the ethics of caring between researcher and researched). These worlds are constituted through a media effect that Hansen terms the *operational present of sensibility* (pp. 195–196), which refers to an increasing distance between perception and sensation, in that we are normally accustomed to perceiving what we sense. For instance, researchers who use microphones for data collection place them into field sites and remain still during the collection itself. They know when they are gathering their data even though they need not sense it or hear it to perceive the recorded sounds. The experience

of sound is thus doubly hidden: microphones at once pick up imperceptible vibrations while keeping those vibrations imperceptible when they
are uploaded onto massive, unlistenable, global databases. Experience and
perception, instead, are now wholly future-oriented, given that one need
not be aware of the environment from which information is being collected. This should sound like an opportunity for a renewed critique of
schizophonia: whereas the twentieth century saw the rise of sound that is
split from the source of its articulation, with twenty-first-century media,
how we perceive sensation is radically decentralized, virtual, and uploaded
on global databases. Sound is now made sense of through algorithms and
code.

The human subject resides in multiple domains at once, through devices
that people in all sorts of professions are using. iPhones are as much used for
recording music and movies as they are for collecting data in research and
checking in on social networks. Instead, Hansen (2015) argues, the human
subject lives in an approximate future anticipation of mediation. In addi-
tion, because these mediations open onto empirical worlds that perception
has no access to otherwise, the "sensory continuum" can encompass the
operational present of sensibility. The subject can live in the future that
unfolds into the past of an unfolding present thereby experiencing several
temporal spans simultaneously.

As mentioned in the introduction, preconvergence sound media
focused more on the object (such as the radio, the piano, the phonograph,
and so on) than on the environment, while current media environments
capture more of a sensory spectrum, as Jonathan Sterne and Mitchell
Akiyama's (2012) plea for more sonification makes clear. And while we
might think this means we need more devices to understand cross-senses,
such as data sonification or sonic visualization, these transductions do not
capture the total image of the sensory spectrum upon which imperceptible
vibrations oscillate. To return to Hansen (2015), he invokes the example
of Etienne-Jules Marey to highlight an instant where the images of motion
held an autonomy of their own (like those of Edward Muybridge's horses):
it is as though they did not just capture a pre-existing reality, but through
their form of capture, contributed a doubled reality. The unnerving try-
pophobia they produce—that of seeing patterns in human movement
that could resemble the Fibonacci series in a pine cone or the perfect

symmetry of a honeycomb—confirms that technology, technique, social organization, form, matter, custom, habit, and movement are organized according to a shared principle. Such principles are determined as much by the instruments that cut into, as represent, these new spaces. Hansen (2015) elaborates:

> What is at issue in Marey's media practice is resolutely not a prosthetic operation of surrogacy, but the veritable inauguration of new, properly technical domains of sensation. In this sense, Marey's machines are the precursors of the complex mechanic networks that supplement human operations today by participating—as autonomous agencies—in the distribution of sensibility beyond perception. Indeed, Marey's conceptualization of machines already grasps everything necessary to understand how today's smart phones, wireless devices, computational micro sensors, and Internet networks interface human experience with new domains to which it lacks direct, perceptual access. (p. 54)

This lack of direct, perceptual access should not close things down, but rather raise occasions to study data as cuts into new realities that demand new questions and new configurations of sense and sensibility. Indeed, while we might be justified in arguing that machines do not sense in the way that humans do, some researchers are recently convinced that machine-sensing is programmable and distributable in a new environmentally programmable digital interface. To reiterate, the new situation, the elemental media buzz on the periphery of human experience does not discredit human experience as a point of interest; instead, it means that human experience, which Hansen maintains is a variant of subjectivity, is multiscalar and occupies human, nonhuman, and techno-human domains.

Subjectivity appears as multiscalar partly because perception and sensation are less intertwined. In contrast, as machine learning and algorithms make possible a mass collection of sensation on the periphery of experience for later analysis and use, human subjectivity is fed forwards to these moments of analysis and reflection. "The resulting scenario," Hansen explains (2015, p. 59), "involves a distinct displacement of our agency: acting through our conscious grasp of situations, we simply cannot have direct operational or real-time access to data milieus of cultural products."

Hanson is not at all concerned that consciousness does not occupy the sensational present. In fact, he looks to the new potential conception of subjectivity as something that can let sensation slip, as potentially unnecessary.

As an example, take musical-error machine learning, which uses algorithms to determine whether music students are performing better or worse from lesson to lesson. With the MIDI console capable of correcting slips in tempo, pitch, and dynamics, a more intimate dynamic is emphasized between the teacher and the student so that those technical slips are subordinated to a context-based performance, meaning that they can be overcome more easily. Delegating the handling of the technical qualities of the performance to a machine means the audience (listeners) can pay more attention to the performer, thus enhancing the interpersonal aspect of the encounter; as listeners, we do not really care about the technical mistakes since they are merely technical trivialities we may not even be aware of.

Throughout this book, new auditory technologies are presented that translate and display the vibrational elements of species and interspecies dynamics. They may enable recordings of new perceptible sounds, but they are not intended as a means to create audio tracings aimed at an enjoyable human listening experience. Instead, listening is now situated on a new ethical continuum between instruments of sense and new research models of global environmental import.

Sounding Worlds

The aim of this book was to open a direction for sound studies that would resemble the shift in media theory theorized by Hansen (2015). Sounding opens onto a field of variation, where any sound powerfully presents across a range of sensory dynamics. This power of variation and repetition makes it impossible to speak of sound as something generic; rather, it is constantly relational and concrete. Sound is the repetition of variations on a singularity that is impossible to generalize, but is linked to its surroundings, successions, and simultaneities. Sounding carries millions of repetitions of millions of vibrations in an infinite direction of relations. Together, these repetitions make rhythm, and it is this rhythm of repeated variations that make up the objects, where "rhythms are the only characters, the only figures" (Deleuze, 2003, p. xv). To invoke a relevant illustrative example from Deleuze, the unfinished and partial bodies in Francis Bacon's *Triptychs* are less representations of bodies than they are figurations of colours, lines, and repetitions in rhythm. Their rhythm, the mutated, stretched bodies, enact a violence of variation. Rhythm is "a vital power that exceeds every domain and traverses them all" (2003, p. 42). To echo Elizabeth Grosz (2008), the power of rhythm traverses any particular form. Rhythm is thus not exclusive to science, music, painting, sport, architecture, or life itself. Rhythm is where sensation meets that vital power in variation and repetition.

Recent trends in media theory have returned to the proposition that the media form determines our experience of the world. While it might be tempting to argue that the voice is a medium of the body, the preceding sketch of sounding shows us that the voice is itself caught up in all sorts of technological and knowledge-based entanglements. Any one history of voicing would demonstrate this point. Thus, following one historical strand (such as voice imaging in the sciences) will not help us arrive at a general definition of sounding. But there is a consistency among the case studies that were explored throughout the book: All soundings begin with a general state of disequilibrium and end on a general interest in vibration. In this crucial sense, sounding is not knowledge of the most anatomical features of the body or the timescales of a place, but instead constitutes knowledge systems about other fields—about violence, technology, science, aesthetics, and social and environmental justice. In this very simple sense, sounding is an interdisciplinary method and requires massively disparate case studies to cohere. Sounding needs a particular agglomeration of technology, organization, and an organ to come alive. And it is also not limited to any one discipline, but to a multiplicity of vibrations and life.

As the brief discussion of the laryngoscope in Chapter 2 demonstrates, sounding opens onto a vibrational field, which is implicated in how space, time, movement, and cycles are ordered and organized in contemporary and historical everyday life. We are much accustomed to associating the voice with vibration, for instance. Voicing is an activated way of thinking through voice. It activates the body, and it activates vibration—voiced in public protest, in production of new soundings, in musical performances, in personal interactions, in research projects.

Thus, must we limit sounding to a human-centric framework? It is certainly caught up in the human possibilities, but as the chapters have demonstrated, it is also tied into elements like light and shadow and fold, movement and flight, vibrational being and the rhythms of place, along with the technical assemblages responsible for its production. Sounding constitutes a purely relational field. A sociologist or a philosopher may approach sound very differently, though every thinker on sound has tried to account for the parallax nature of sounding: that it is at once a physical but symbolic presence, physically present but simultaneously its own representation.

This book is meant to join the move beyond laryngealcentric and body/subjectivity-focused notions of sound by developing the concept of sounding. This challenge is not an isolated one, but rather is one of many contemporary challenges to body/subjectivity that uses voice, body, sound, and vibration. While sounding has been the guiding theme throughout this book, we have also benefited from the insights of Stiegler's general organology (Chapter 2); Deleuze and Guattari's incorporeal transformation (Chapter 3); Barry Truax's acoustic community (Chapter 4); Timothy Morton's hyperobject (Chapter 6); and the ideas of other contemporary scholars such as Mark Hansen, Jennifer Gabrys, and Stefan Helmreich. Together, they have helped articulate a philosophy of sound and sounding that moves beyond the cochlear and laryngeal, and towards the vibrational and the ethical, governance and the earth. Collectively, their work makes room for a bio/human/nature cluster that is networked, digital, natural, corporeal, and live. With an eye to contributing to a posthuman ethic through the concept of sounding, I hope that future work contributes further to the ongoing transition in media theory from media objects to media environments, by tracing sound objects towards sonic immersion, worlds, and ecologies.

References

Bijsterveld, K. (2019). *Sonic skills: Listening for knowledge in science, medicine and engineering (1920s–present)*. London: Palgrave Macmillan.

Bruyninckx, J. (2018). *Listening in the field: Recording and the science of birdsong*. Cambridge: MIT Press.

Deleuze, G. (2003). *Francis Bacon: The logic of sensation*. London: Continuum.

Deleuze, G., & Guattari, F. (1987). *A thousand plateaus: Capitalism and schizophrenia* (B. Massumi, Trans.). Minneapolis: University of Minnesota Press.

Friedner, M., & Helmreich, S. (2012). Sound studies meets deaf studies. *The Senses and Society, 7*(1), 72–86.

Gabrys, J. (2016a). *Program earth: Environmental sensing technology and the making of a computational planet*. Minneapolis: University of Minnesota Press.

Gabrys, J. (2016b). Programming environments: Environmentality and citizen sensing in the smart city. *Environment and Planning D: Society and Space, 32*, 30–48.

Grosz, E. (2008). *Chaos, territory, art: Deleuze and the framing of the earth*. New York: Columbia University Press.

Hansen, M. (2015). *Feed-forward: On the future of twenty-first media*. Chicago: University of Chicago Press.

Henriques, J. (2008). Sonic diaspora, vibrations, and rhythm: Thinking through the sounding of the Jamaican dancehall session. *African and Black Diaspora: an International Journal, 1*(2), 215–236.

Ross, A. (2012). Song of the earth. In B. Herzogenrath (Ed.), *The farthest place: The music of John Luther Adams* (pp. 13–22). Princeton, NJ: Northeastern University Press.

Schulze, H. (2018). *The sonic persona: An anthropology of sound*. New York: Bloomsbury Academic.

Smith, D. N. (2012). *Sounding/silence: Martin Heidegger and the limits of poetics*. New York: Fordham University Press.

Sterne, J., & Akiyama, M. (2012). The recording that never wanted to be heard and other stories of sonification. In T. Pinch & K. Bijsterveld (Eds.), *The Oxford handbook of sound studies* (pp. 544–560). Oxford: Oxford University Press.

Supper, A. (2014). Sublime frequencies: The construction of sublime listening experiences in the sonification of scientific data. *Social Studies of Science, 44*(1), 34–58.

References

Adams, J. L. (2010). *The place where you go to listen: In search of an ecology of music*. Middletown, CT: Wesleyan University Press.

Adams, M., Cox, T., Moore, G., Croxford, B., Refaee, M., & Sharples, S. (2006). Sustainable soundscapes: Noise policy and the urban experience. *Urban Studies, 43*(13), 2385–2398.

Agamben, G. (2004). *The open: Man and animal*. Palo Alto, CA: Stanford University Press.

Alberta Culture and Tourism. (2016). *Frank Slide Interpretive Centre: FAQ— Frequently asked questions*. Retrieved from http://www.history.alberta.ca/frankslide/faq/faq.aspx.

Alberta Energy Regulator. (2016a). *Turtle Mountain monitoring program*. Retrieved from http://ags.aer.ca/turtle-mountain-monitoring-program.htm.

Alberta Energy Regulator. (2016b). *Historical monitoring at Turtle Mountain*. Retrieved from http://ags.aer.ca/historical-monitoring-at-turtle-mountain.

Alberta Energy Regulator. (2016c). *Alberta geological survey: Current monitoring*. Retrieved from http://ags.aer.ca/current-monitoring.

Alexander, D. (1991). Information technology in real-time for monitoring and managing natural disasters. *Progress in Physical Geography, 15*, 238–260.

Ali, Z., Muhammad, G., & Alhamid, M. F. (2017). An automatic health monitoring system for patients suffering from voice complications in smart cities. *IEEE Access, 5*, 3900–3908.

© The Editor(s) (if applicable) and The Author(s) 2020
M. Vallee, *Sounding Bodies Sounding Worlds*, Palgrave Studies in Sound,
https://doi.org/10.1007/978-981-32-9327-4

Allen, M. (2010). VoxNet: Reducing latency in high data rate applications. In E. Gaura, L. Girod, J. Brusey, M. Allen, & G. Challen (Eds.), *Wireless sensor networks: Deployments and design frameworks* (pp. 115–158). London, UK: Springer.

Andersen, J. (2015). Now you've got the shiveries: Affect, intimacy, and the ASMR whisper community. *Television & New Media, 16*(8), 683–700.

Ang, I. (2006). From cultural studies to cultural research. *Cultural Studies Review, 12,* 183–197.

Antick, P. (2013). Bhopal to Bridgehampton: Schema for a disaster tourism event. *Journal of Visual Culture, 12,* 164–185.

Aronson, A. E., & Bless, D. M. (2011). *Clinical voice disorders* (4th ed.). New York, NY: Thieme Publishing.

Asdal, K. (2008). Subjected to parliament: The laboratory of experimental medicine and the animal body. *Social Studies of Science, 38*(6), 899–917.

August, T., Harvey, M., Lightfoot, P., Kilbey, D., Papadopoulos, T., & Jepson, P. (2015). Emerging technologies for biological recording. *Biological Journal of the Linnean Societ, 115*(3), 731–749.

Back, L. (2007). *The art of listening.* London, UK: Berg.

Bailey, B. (1996). Laryngoscopy and laryngoscopes—Who's first? The forefathers/four fathers of laryngology. *The Laryngoscope, 106*(8), 939–943.

Baptista, L. F., & Keister, R. A. (2005). Why birdsong is sometimes like music. *Perspectives in Biology and Medicine, 48*(3), 426–443.

Baron, B. C., & Dedo, H. H. (1980). Separation of the larynx and trachea for intractable aspiration. *The Laryngoscope, 90*(12), 1927–1932.

Barthes, R. (1977). The grain of the voice (S. Heath, Trans.). In R. Barthes (Ed.), *Image-music-text* (pp. 179–189). London, UK: Fontana.

Bayne, E. M., Habib, L., & Boutin, S. (2008). Impacts of chronic anthropogenic noise from energy sector activity on abundance of songbirds in the boreal forest. *Conservation Biology, 22*(5), 1186–1193.

Benko, B., & Stead, D. (1998). The frank slide: A reexamination of the failure mechanism. *Canadian Geotechnical Journal, 35,* 299–311.

Benschop, R. (2007). Memory machines or musical instruments? Soundscapes, recording technologies and reference. *International Journal of Cultural Studies, 10*(4), 485–502.

Berger, E. (2015). Welcome to the quietest square inch in the U.S. outside. *Outside Online.* Retrieved from https://www.outsideonline.com/2000721/welcome-quietest-square-inch-us.

Berland, J. (2009). *North of empire: Essays on the cultural technologies of space.* Durham: Duke University Press.

Berlant, L. (2011). *Cruel optimism.* Durham, NC: Duke University Press.

Bernstein, C. (2009). Making audio visible: The lessons of visual language for the textualization of sound. *Textual Practice, 23*(6), 959–973.

Bernstein, M. H. (1998). *On moral considerability: An essay on who morally matters.* Oxford, UK: Oxford University Press.

Bijsterveld, K. (2019). *Sonic skills: Listening for knowledge in science, medicine and engineering (1920s–present).* London: Palgrave Macmillan.

Blackman, L. (2000). Ethics, embodiment and the voice-hearing experience. *Theory, Culture & Society, 17*(5), 55–74.

Blackman, L. (2010). Embodying affect: Voice-hearing, telepathy, suggestion and modelling the non-conscious. *Body & Society, 16*(1), 163–192.

Blackman, L. (2012). *Immaterial bodies: Affect, embodiment, mediation.* London, UK: Sage.

Blackman, L. (2016). The challenges of new biopsychosocialities: Hearing voices, trauma, epigenetics and mediated perception. *Sociological Review Monographs, 64*(1), 256–273.

Blesser, B., & Salter, L. R. (2007). *Spaces speak, are you listening? Experiencing aural architecture.* Cambridge: MIT Press.

Blumstein, D. T., Mennill, D. J., Clemins, P., Girod, L., Yao, K., Patricelli, G., … Kirschel, A. (2011). Acoustic monitoring in terrestrial environments using microphone arrays: Applications, technological considerations and prospectus. *Journal of Applied Ecology, 48*(3), 758–767.

Bode, R. (2014). Rhythm and its importance for education. *Body & Society, 20*(3–4), 51–74.

Bolhuis, J. J., & Wynne, C. D. L. (2009). Can evolution explain how minds work? *Nature, 458,* 832–833.

Bonnard, C. (2011). Technical and human aspects of historic rockslide-dammed lakes and landslide dam breaches. In S. G. Evans, R. L. Hermanns, A. Strom, & G. Scarascia-Mugnozza (Eds.), *Natural and artificial rockslide dams* (pp. 101–122). Berlin, Germany: Springer.

Borker, A. L., Halbert, P., McKown, M. W., Tershy, B. R., & Croll, D. A. (2015). A comparison of automated and traditional monitoring techniques for marbled murrelets using passive acoustic sensors. *Wildlife Society Bulletin, 39*(4), 813–818.

Bourdieu, P. (1984). *Distinction: A social critique of the judgement of taste.* Cambridge, MA: Harvard University Press.

Bowman, M. S., & Pezzullo, P. C. (2010). What's so "dark" about "dark tourism"? Death, tours, and performance. *Tourist Studies, 9,* 187–202.

Bozzini, P. (1807). *Der Lichtleiter oder Beschreibung einer einfachen Vorrichtung und ihrer Anwendung zur Erleuchtung innerer Höhlen und Zwischenräume des lebenden animalischen Körpers.* Landes-Industrie-Comptoir.

Brand, A. R. (1937). Why bird song cannot be described adequately. *The Wilson Bulletin, 49*(1), 11–14.

Brentari, C. (2015). *Jakob von Uexküll: The discovery of the umwelt between biosemiotics and theoretical biology.* New York, NY: Springer.

Bromhead, E. N., Canuti, P., & Ibsen, M.-L. (2006). Landslides and cultural heritage. *Landslides, 3,* 273–274.

Bruyninckx, J. (2012). Sound sterile: Making scientific field recordings in ornithology. In T. Pinch & K. Bijsterveld (Eds.), *The Oxford handbook of sound studies* (pp. 127–150). Oxford, UK: Oxford University Press.

Bruyninckx, J. (2015). Trading twitter: Amateur recorders and economies of scientific exchange at the Cornell Library of Natural Sounds. *Social Studies of Science, 45*(3), 344–370.

Bruyninckx, J. (2018). *Listening in the field: Recording and the science of birdsong.* Cambridge: MIT Press.

Buller, H. (2013). Individuation, the mass and farm animals. *Theory, Culture & Society, 30*(7–8), 155–175.

Butler, J. (1997). *Excitable speech: A politics of the performative.* London, UK: Routledge.

Cage, J. (2016). To Mario Cavista. In L. Kuhn (Ed.), *The selected letters of John Cage* (p. 477). Middletown, CT: Wesleyan University Press.

Candea, M. (2013). Habituating meerkats and redescribing animal behaviour science. *Theory, Culture & Society, 30*(7–8), 105–128.

Canuti, P., Margottini, C., Fanti, R., & Bromhead, E. N. (2009). Cultural heritage and landslides: Research for risk prevention and conservation. In K. Sassa & P. Canuti (Eds.), *Landslides: Disaster risk reduction* (pp. 401–433). Berlin, Germany: Springer.

Cavarero, A. (2005). *For more than one voice: Toward a philosophy of vocal expression* (P. A. Kottman, Trans.). Stanford, CA: Stanford University Press.

Cepstral: We build voices. Retrieved from http://www.cepstral.com/en/demos. Accessed September 7, 2015.

Chandola, T. (2012). Listening in to water routes: Soundscapes as cultural systems. *International Journal of Cultural Studies, 16*(1), 55–69.

Chandrasekera, T., Yoon, S.-Y., & D'Souza, N. (2015). Virtual environments with soundscapes: A study on immersion and effects of spatial abilities. *Environment and Planning B: Planning and Design, 42,* 1003–1019.

Chang Hwan, R., Seung, H., Moo-Song, L., Sang Yoon, K., Soon Yuhl, N., John-Lyel, R., ... Seung-Ho, C. (2015). Voice changes in elderly adults: Prevalence and the effect of social, behavioral, and health status on voice quality. *Journal of the American Geriatrics Society.* Retrieved from https://doi.org/10.1111/jgs.13559.

Charrière, M., Humair, F., Froese, C., Jaboyedoff, M., Pedrazzini, A., & Longchamp, C. (2016). From the source area to the deposit: Collapse, fragmentation, and propagation of the Frank Slide. *Geological Society of America Bulletin, 128,* 332–351.

Chatziprokopiou, M. (2015). Lamenting (with the) "others", "lamenting our failure to lament"? An auto-ethnographic account of the vocal expression of loss. In K. Thomaidis & B. Macpherson (Eds.), *Voice studies: Critical approaches to process, performance and experience* (pp. 149–161). London, UK: Routledge.

Chiew, F. (2014). Posthuman ethics with Cary Wolfe and Karen Barad: Animal compassion as trans-species entanglement. *Theory, Culture & Society, 31*(4), 51–69.

Chion, M. (1999). *The voice in cinema* (C. Gorbman, Trans.). New York, NY: Columbia University Press.

Clark, N. (2013). Geoengineering and geologic politics. *Environment and Planning A, 45,* 2825–2832.

Clough, P. (2009). The new empiricism: Affect and sociological method. *European Journal of Social Theory, 12*(1), 43–61.

Clough, P. (2010). Afterword: The future of affect studies. *Body & Society, 16*(1), 222–230.

Cohn, J. P. (2008). Citizen science: Can volunteers do real research? *AIBS Bulletin, 58*(3), 192–197.

Connor, S. (2000). *Dumbstruck: A cultural history of ventriloquism.* Oxford, UK: Oxford University Press.

Connors, J. P., Patrick, J., Lei, S., & Kelly, M. (2012). Citizen science in the age of neogeography: Utilizing volunteered geographic information for environmental monitoring. *Annals of the Association of American Geographers, 102*(6), 1267–1289.

Conrad, P. (2007). *The medicalization of society: On the transformation of human conditions into treatable disorders.* Baltimore, MD: Johns Hopkins University Press.

Corey, G. (2009). *Now this is a site to see.* Retrieved from https://www.yelp.ca/biz/the-frank-slide-crowsnest-pass.

Cornwell, M. L., & Campbell, L. M. (2012). Co-producing conservation and knowledge: Citizen-based sea turtle monitoring in North Carolina, USA. *Social Studies of Science, 42*(1), 101–120.

Crespi, P. (2014). Rhythmanalysis in gymnastics and dance: Rudolf Bode and Rudolf Laban. *Body & Society, 20*(3&4), 30–50.

Crist, E. (1996). Naturalists' portrayals of animal life: Engaging the verstehen approach. *Social Studies of Science, 26*(4), 799–838.

Crist, E. (2004). Can an insect speak? The case of the honeybee dance language. *Social Studies of Science, 34*(1), 7–43.

Cruden, D. M., & Martin, C. D. (2007). Before the Frank Slide. *Canadian Geotechnical Journal, 44,* 765–780.

Cruden, D., & Lan, H. X. (2015). Using the working classification of landslides to assess the danger from a natural slope. In G. Lollino, A. Manconi, J. Clague, W. Shan, & M. Chiarle (Eds.), *Engineering geology for society and territory* (Vol. 2, pp. 3–12). Basel, Switzerland: Springer.

Currie, P. J. (1993). Black beauty. *Dino Frontline, 4,* 22–36.

Czermak, J. N. (1861). On the laryngoscope and its employment in physiology and medicine. *New Sydenham Society, 11,* 1–79.

De Coensel, B., & Botteldooren, D. (2007). The rhythm of the urban sound-scape. *Noise and Vibration Worldwide, 38*(9), 11–17.

Degnen, C. (2005). Relationality, place, and absence: A three-dimensional perspective on social memory. *Sociological Review, 53,* 729–744.

de la Bellacasa, M. P. (2017). *Matters of care: Speculative ethics in more than human worlds.* Minneapolis: University of Minnesota Press.

Deleuze, G. (1986). *Cinema 1: The movement image.* London: A & C Black.

Deleuze, G. (1988). *Spinoza: Practical philosophy* (R. Hurley, Trans.). San Francisco, CA: City Lights Books.

Deleuze, G. (1989). *Cinema II: The time-image.* London, UK: Continuum.

Deleuze, G. (2003). *Francis Bacon: The logic of sensation.* London: Continuum.

Deleuze, G., & Guattari, F. (1986). *Kafka: Toward a minor literature.* Minneapolis: University of Minnesota Press.

Deleuze, G., & Guattari, F. (1987). *A thousand plateaus: Capitalism and schizophrenia* (B. Massumi, Trans.). Minneapolis: University of Minnesota Press.

DeLyser, D., & Sui, D. (2013). Crossing the qualitative-quantitative divide II: Inventive approaches to big data, mobile methods, and rhythmanalysis. *Progress in Human Geography, 37*(2), 293–305.

de Sagazan, O. (2012). *Transfiguration.* Angers, France: Presses de l'université d'Angers.

Despret, V. (2013). Responding bodies and partial affinities in human–animal worlds. *Theory, Culture & Society, 30*(7/8), 66–91.

Di Matteo, P. (2015a). Capture of speech in (dis)embodied voices. In K. Thomaidis & B. Macpherson (Eds.), *Voice studies: Critical approaches to process, performance and experience* (pp. 104–119). London, UK: Routledge.

Di Matteo, P. (2015b). Performing the entre-deux: The capture of speech in (dis)embodied voices. In K. Thomaidis & B. Macpherson (Eds.), *Voice studies: Critical approaches to process, performance and experience* (pp. 90–103). London, UK: Routledge.

Dolar, M. (2006). *A voice and nothing more.* Cambridge: MIT Press.

Donaldson, A. (2016, July 11). National network of acoustic recorders proposed to eavesdrop on australian ecosystems. *ABC News.* Retrieved from http://www.abc.net.au/news/2016-07-11/soundscape-ecology-could-track-environmental-changes/7587354.

Dyson, F. (2009). *Sounding new media: Immersion and embodiment in the arts and culture.* London, UK: University of California Press.

Echternach, M. (2016). Magnetic resonance imaging of the voice production system. In R. T. Sataloff (Ed.), *Professional voice: The science and art of clinical care* (4th ed.). San Diego, CA: Plural Publishing.

Eidsheim, N. (2012). Voice as action: Toward a model for analyzing the dynamic construction of racialized voice. *Current Musicology, 93,* 9–33.

Eidsheim, N. (2014). The micropolitics of listening to vocal timbre. *Postmodern Culture, 24*(3). Retrieved from https://doi.org/10.1353/pmc.2014.0014.

Eidsheim, N. S. (2015). *Sensing sound: Singing and listening as vibrational practice.* Durham, NC: Duke University Press.

Eidsheim, N. S., & Mazzei, L. A. (2019). *The Oxford handbook of voice studies.* Oxford, UK: Oxford University Press.

Ekström, A. (2017). Remediation, time and disaster. *Theory, Culture & Society, 33*(5), 117–138.

Farina, A. (2014). *Soundscape ecology: Principles, patterns, methods and applications.* London, UK: Springer.

Farina, A., Lattanzi, E., Malavasi, R., Pieretti, N., & Piccioli, L. (2011). Avian soundscapes and cognitive landscapes: Theory, application and ecological perspectives. *Landscape Ecology, 26*(9), 1257–1267.

Farina, A., Pieretti, N., & Piccioli, L. (2011). The soundscape methodology for long-term bird monitoring: A Mediterranean Europe case-study. *Ecological Informatics, 6*(6), 354–363.

Fitch, W. T. (2000). The evolution of speech: A comparative review. *Trends in Cognitive Sciences, 4*(7), 258–267.

Fong, J. (2016). Making operative concepts from Murray Schafer's soundscapes typology: A qualitative and comparative analysis of noise pollution in Bangkok, Thailand, and Los Angeles, California. *Urban Studies, 53*(1), 173–192.

Fraser, C. S., & Gruendig, L. (1985). The analysis of photogrammetric deformation measurements on Turtle Mountain. *Photogrammetric Engineering and Remote Sensing, 51,* 207–216.

Friedner, M., & Helmreich, S. (2012). Sound studies meets deaf studies. *The Senses and Society, 7*(1), 72–86.

Friese, C., & Clarke, A. E. (2012). Transposing bodies of knowledge and technique: Animal models at work in reproductive sciences. *Social Studies of Science, 42*(1), 31–52.

Gabrys, J. (2016a). *Program earth: Environmental sensing technology and the making of a computational planet.* Minneapolis: University of Minnesota Press.

Gabrys, J. (2016b). Programming environments: Environmentality and citizen sensing in the smart city. *Environment and Planning D: Society and Space, 32,* 30–48.

Gagliano, M., Mancuso, S., & Robert, D. (2012). Towards understanding plant bioacoustics. *Trends in Plant Science, 17*(6), 323–325.

Gallagher, M. (2015a). Field recording and the sounding of spaces. *Environment and Planning D: Society and Space, 33*(3), 560–576.

Gallagher, M. (2015b). Sounding ruins: Reflections on the production of an "audio drift." *Cultural Geographies, 22*(3), 467–485.

Gallagher, M., Kanngieser, A., & Prior, J. (2017). Listening geographies: Landscape, affect and geotechnologies. *Progress in Human Geography, 41*(5), 618–637.

García, M. (1855). Observations on the human voice (Proceedings of learned societies). *Philosophical Magazine, 10*(65), 217–229.

García, M. (1881). On the invention of the laryngoscope. *Transactions of the international medical congress, seventh session* (pp. 197–199). London, UK: International Medical Congress.

Gardiner, M. E. (2012). Henri Lefebvre and the 'sociology of boredom'. *Theory, Culture & Society, 29*(2), 37–62.

Gill, L. F., D'Amelio, P. B., Adreani, N. M., Sagunsky, H., Gahr, M. C., & ter Maat, A. (2016). A minimum-impact, flexible tool to study vocal communication of small animals with precise individual-level resolution. *Methods in Ecology and Evolution, 7*(11), 1349–1358.

Giraud, E., & Hollin, G. (2016). Care, laboratory beagles and affective utopia. *Theory, Culture and Society, 33*(4), 27–49.

Glass, K. (2017). *Politics and affect in black women's fiction.* London: Lexington Books.

Goodman, S. (2010). *Sonic warfare: Sound, affect, and the ecology of fear.* Cambridge: MIT Press.

Gough, K. M. (2012). Natural disaster, cultural memory: Montserrat adrift in the black and green Atlantic. In W. Arons & T. J. May (Eds.), *Readings in performance and ecology* (pp. 101–112). New York, NY: Palgrave Macmillan.

Government of Alberta, Alberta Emergency Management Agency. (2017). *Emergency response protocol for Turtle Mountain.* Retrieved from http://www.aema.alberta.ca/documents/Emergency-Response-Protocol-for-Turtle-Mountain.pdf.

Greenwood, J. J. D. (2007). Citizens, science and bird conservation. *Journal of Ornithology, 148*(Suppl. 1), S77–S124.

Grimshaw, M., & Garner, T. (2015). *Sonic virtuality: Sound as emergent perception.* Oxford, UK: Oxford University Press.

Grosz, E. (2008). *Chaos, territory, art: Deleuze and the framing of the earth.* New York: Columbia University Press.

Guggenheim, M. (2014). Introduction: Disasters as politics–politics as disasters. *Sociological Review, 62*(Suppl. 1), 1–16.

Gvion, L. (2016). "If you ever saw an opera singer naked": The social construction of the singer's body. *European Journal of Cultural Studies, 19*(2), 150–169.

Haggard, P. (2008). Human volition: Towards a neuroscience of will. *Nature Reviews Neuroscience, 9,* 934–946.

Hall, M. (2016). Soundscape ecology: Eavesdropping on nature. *Deutsche Well (DW).* Retrieved from http://www.dw.com/en/soundscape-ecology-eavesdropping-on-nature/a-19304871.

Hall, T., Lashua, B., & Coffey, A. (2008). Sound and the everyday in qualitative research. *Qualitative Inquiry, 14*(6), 1019–1040.

Hansen, M. (2015). *Feed-forward: On the future of twenty-first media.* Chicago: University of Chicago Press.

Haraway, D. (2008). *When species meet.* Minneapolis: University of Minnesota Press.

Harrison, T. W., & Friesen, J. W. (2015). *Canadian society in the twenty-first century: An historical sociological approach* (3rd ed.). Toronto, ON, Canada: Canadian Scholars' Press.

Haug, Ø. T., Rosenau, M., Leever, K., & Oncken, O. (2016). On the energy budgets of fragmenting rockfalls and rockslides: Insights from experiments. *Journal of Geophysical Research: Earth Surface, 121,* 1310–1327.

Hawkins, D. (2011). "Soundscape ecology": The new science helping identify ecosystems at risk. *Ecologist: Setting the environmental agenda since 1970.* Retrieved from https://theecologist.org/investigations/science_and_technology/1171165/soundscape_ecology_the_new_science_helping_identify_ecosystems_at_risk.html.%20Accessed%20September%2030,%202017.

Helmholtz, H. L. F. (1885). *On the sensations of tone as a physiological basis for the theory of music* (A. J. Ellis, Trans.). London, UK: Longmans, Green.

Helmreich, S. (2015). *Sounding the limits of life: Essays in the anthropology of biology and beyond.* Princeton: Princeton University Press.

Henriques, J. (2008). Sonic diaspora, vibrations, and rhythm: Thinking through the sounding of the Jamaican dancehall session. *African and Black Diaspora: An International Journal, 1*(2), 215–236.

Henriques, J. (2010). The vibrations of affect and their propagation on a night out on Kingston's dancehall scene. *Body & Society, 16*(1), 57–89.

Henriques, J. (2011). *Sonic bodies: Reggae sound systems, performance techniques, and ways of knowing.* London, UK: Bloomsbury.

Henriques, J. (2014). Rhythmic bodies: Amplification, inflection and transduction in the dance performance techniques of the "Bashment gal". *Body & Society, 20*(3–4), 79–112.

Herbst, C. T., Stoeger, A. S., Frey, R., Lohscheller, J., Titze, I. R., Gumpenberger, M., & Fitch, W. T. (2012). How low can you go? Physical production mechanism of elephant infrasonic vocalizations. *Science, 337*(6094), 595–599.

Hernández-Moreno, G., & Alcántara-Ayala, I. (2016). Landslide risk perception in Mexico: A research gate into public awareness and knowledge. *Landslides, 14*(1). Retrieved from https://doi.org/10.1007/s10346-016-0683-9.

Hetherington, K. (2003). Spatial textures: Place, touch, and praesentia. *Environment and planning A, 35*(11), 1933–1944.

Hill, P. (2007). *Olivier messiaen: Oiseaux exotiques.* Burlington, VT: Ashgate.

Hossain, M. S., Muhammad, G., & Alamri, A. (2017). Smart healthcare monitoring: A voice pathology detection paradigm for smart cities. *Multimedia Systems.* Retrieved from https://doi.org/10.1007/s00530-017-0561-x.

Humair, F., Pedrazzini, A., Epard, J. L., Froese, C. R., & Jaboyedoff, M. (2013). Structural characterization of Turtle Mountain anticline (Alberta, Canada) and impact on rock slope failure. *Tectonophysics, 605,* 133–148.

Hutto, R. L., & Stutzman, R. J. (2009). Humans versus autonomous recording units: A comparison of point-count results. *Journal of Field Ornithology, 80*(4), 387–398.

Ikoniadou, E. (2014). Abstract time and affective perception in the sonic work of art. *Body & Society, 20*(3&4), 140–161.

International Bioacoustics Council. (2019). http://www.ibac.info.

Irwin, A. (1995). *Citizen science: A study of people, expertise and sustainable development.* London: Routledge.

Jackson, A. Y., & Mazzei, L. (2011). *Thinking with theory in qualitative research: Viewing data across multiple perspectives.* London: Routledge.

James, P. (1885). *Laryngoscopy and rhinoscopy in the diagnosis and treatment of diseases of the throat and nose.* London, UK: Baillière, Tindall, and Cox.

Järviö, P. (2015). The singularity of experience in the voice studio: A dialogue with Michael Henry. In K. Thomaidis & B. Macpherson (Eds.), *Voice studies: Critical approaches to process, performance and experience* (pp. 25–37). London, UK: Routledge.

Johnson, J. J. (1995). *Listening in Paris: A cultural history.* Berkeley, CA: University of California Press.

Kahn, D. (1999). *Noise, water, meat: A history of sound in the arts.* Cambridge: MIT Press.

Kanngieser, A. (2012). A sonic geography of voice: Towards an affective politics. *Progress in Human Geography, 36*(3), 336–353.

Katz, M. (2010). *Capturing sound: How technology has changed music.* Berkeley: University of California Press.

Kircher, A. (1650/1970). *Musurgia universalis.* Hildesheim and New York: G. Olms (Reprint of the Rome, 1650 edition).

Kitchin, R. (2014). *The data revolution: Big data, open data, data infrastructures, and their consequences.* London, UK: Sage.

Korstanje, M. E. (2016). *The rise of thana-capitalism and tourism.* London: Routledge.

Kronenberg, J. (2014). Environmental impacts of the use of ecosystem services: Case study of birdwatching. *Environmental Management, 54*(3), 617–630.

Kuhnke, E. (2012). *Persuasion and influence for dummies.* Hoboken, NJ: Wiley.

La Barbara, J. (2002). Voice is the original instrument. *Contemporary Music Review, 21*(1), 35–48.

LaBelle, B. (2014). *Lexicon of the mouth: Poetics and politics of voice and the oral imaginary.* London, UK: Bloomsbury.

Laiolo, P. (2010). The emerging significance of bioacoustics in animal species conservation. *Biological Conservation, 143*(7), 1635–1645.

Lameira, A. R., Hardus, M. E., Mielke, A., Wich, S. A., & Shumaker, R. W. (2016). Vocal fold control beyond the species-specific repertoire in an orangutan. *Scientific Reports, 6.* Retrieved from https://www.nature.com/articles/srep30315.

Larkin, B. (2013). The politics and poetics of infrastructure. *Annual Review of Anthropology, 42,* 327–343.

Latimer, J. (2013). Being alongside: Rethinking relations among different kinds. *Theory, Culture & Society, 30*(7/8), 77–104.

Latour, B. (1999). *Pandora's hope: Essays on the reality of science studies.* Cambridge, MA: Harvard University Press.

Latour, B. (2005). *Reassembling the social: An introduction to actor-network-theory.* Oxford: Oxford University Press.

Lawrence, A. (2006). "No personal motive?" Volunteers, biodiversity, and the false dichotomies of participation. *Ethics, Place and Environment, 9*(3), 279–298.

Lefebvre, H. (2004). *Rhythmanalysis: Space, time and everyday life.* London: A&C Black.

Lennon, J. J., & Foley, M. (2000). *Dark tourism.* London: Cengage Learning EMEA.

Levitt, R. (2013). Silence speaks volumes: Counter-hegemonic silences, deafness, and alliance work. In S. Malhotra & A. E. Rowe (Eds.), *Silence, feminism, power: Reflections at the edges of sound* (pp. 67–83). London, UK: Palgrave Macmillan.

Lievrouw, L. A. (2010). Social media and the production of knowledge: A return to little science? *Social Epistemology, 24*(3), 219–237.

Ling, C., Li, Q., Brown, M. E., Kishimoto, Y., Toya, Y., Devine, E. E., & Welman, N. V. (2015). Bioengineered vocal fold mucosa for voice restoration. *Science Translational Medicine, 7*(314). Retrieved from https://stm.sciencemag.org/content/7/314/314ra187.

Lorimer, J. (2012). Multinatural geographies for the Anthropocene. *Progress in Human Geography, 36*(5), 593–612.

Lury, C., Parisi, L., & Terranova, T. (2012). Introduction: The becoming topological of culture. *Theory, Culture & Society, 29*(4–5), 3–35.

Lynch, M. E. (1988). Sacrifice and the transformation of the animal body into a scientific object: Laboratory culture and ritual practice in the neurosciences. *Social Studies of Science, 18*(2), 265–289.

Mair, M., Greiffenhagen, C., & Sharrock, W. W. (2015). Statistical practice: Putting society on display. *Theory, Culture & Society, 33*(3), 51–77.

Malloch, S. E., & Trevarthen, C. E. (Eds.). (2009). *Communicative musicality: Exploring the basis of human companionship.* Oxford, UK: Oxford University Press.

Malone, N., Wade, A. H., Fuentes, A., Riley, E. P., Remis, M., & Robinson, C. J. (2014). Ethnoprimatology: Critical interdisciplinarity and multispecies approaches in anthropology. *Critique of Anthropology, 34*(1), 8–29.

Manning, E. (2014). Wondering the world directly—Or, how movement outruns the subject. *Body & Society, 20*(3–4), 162–188.

Manovich, L. (2001). *The language of new media.* Cambridge: MIT Press.

Marty, S. (2011). *Leaning on the wind: Under the spell of the Great Chinook.* Surrey, BC, Canada: Heritage.

Massey, D. (2005). *For space.* London: Sage.

Massumi, B. (2002). *Parables for the virtual: Movement, affect, sensation.* Durham: Duke University Press.

Mazumder, A. (2016). *Pacific North West LNG Project: A review and assessment of the project plans and their potential impacts on marine fish and fish habitat in the Skeena estuary.* Environmental Assessment Report, Government of Canada. Minister of Environment and Climate Change. https://www.ceaa.gc.ca/050/evaluations/proj/80032?culture=en-CA.

Mazzei, L. A. (2009). An impossibly full voice. In A. Y. Jackson & L. A. Mazzei (Eds.), *Voice in qualitative inquiry: Challenging conventional, interpretive, and critical conceptions in qualitative research* (pp. 45–62). London, UK: Routledge.

Mazzei, L. A. (2013). A voice without organs: Interviewing in posthumanist research. *International Journal of Qualitative Studies in Education, 26*(6), 732–740.

Mazzei, L. A., & Jackson, A. Y. (2012). Complicating voice in a refusal to "let participants speak for themselves". *Qualitative Inquiry, 18*(9), 745–751.

Mazzei, L. A., & McCoy, K. (2010). Thinking with Deleuze in qualitative research. *International Journal of Qualitative Studies in Education, 23*(5), 503–509.

McConnell, R. G., & Brock, R. W. (1904). *Report on the great landslide at Frank, Alta. 1903: Extract from part VIII, annual report, 1903.* Ottawa, ON, Canada: Government Printing Bureau.

McLean, E. (2016). Exploring parents' experiences and perceptions of singing and using their voice with their baby in a neonatal unit: An interpretative phenomenological analysis. *Article in Nordic Journal of Music Therapy, 11,* 1–42.

Meloni, M. (2014). How biology became social, and what it means for social theory. *The Sociological Review, 62,* 593–614.

Melosh, H. J. (1986). The physics of very large landslides. *Acta Mechanica, 64*(1–2), 89–99.

Merchant, N. D., Fristrup, K. M., Johnson, M. P., Tyack, P. L., Witt, M. J., Blondel, P., & Parks, S. E. (2015). Methods in ecology and evolution. *Methods in Ecology and Evolution, 6*(3), 257–265.

Merkur.de. (2016). *MRT-Aufnahmen von Michael Volle bei "Lied an den Abendstern"* [Gruseliges video]. Retrieved from https://www.youtube.com/watch?v=GCluRCd2YuM.

Merleau-Ponty, M. (1962). *The phenomenology of perception* (C. Smith, Trans.). London, UK: Routledge & Kegan Paul.

Merleau-Ponty, M. (1968). *The visible and the invisible: Followed by working notes* (A. Lingis, Trans.). Evanston, IL: Northwestern University Press.

Mesaros, A., Heittola, T., & Virtanen, T. (2016, August). TUT database for acoustic scene classification and sound event detection. In *2016 24th European Signal Processing Conference (EUSIPCO)* (pp. 1128-1132). IEEE.

Michael, M. (2012). Anecdote. In C. Lury & N. Wakeford (Eds.), *Inventive methods: The happening of the social* (pp. 25–35). London, UK: Routledge.

Michael, M., & Birke, L. (1994). Enrolling the core set: The case of the animal experimentation controversy. *Social Studies of Science, 24*(1), 81–95.

Miller, D. (2006). *Be heard for the first time: The woman's guide to powerful speaking.* Sterling, MD: Capital Books.

Moritz, C., Patton, J. L., Conroy, C. J., Parra, J. L., White, G. C., & Beissinger, S. R. (2008). Impact of a century of climate change on small-mammal communities in Yosemite National Park, USA. *Science, 322,* 261–264.

Morton, T. (2010). *The ecological thought.* Cambridge, MA: Harvard University Press.

Morton, T. (2013). *Hyperobjects: Philosophy and ecology after the end of the world.* Minneapolis: University of Minnesota Press.

Mrázek, J. (2015). A sea of honey: The speaking voice in Javanese shadow-puppet theatre. In K. Thomaidis & B. Macpherson (Eds.), *Voice studies: Critical approaches to process, performance and experience* (pp. 64–76). London, UK: Routledge.

Mundy, R. (2009). Birdsong and the image of evolution. *Society and Animals, 17,* 206–223.

Nelson, N. C. (2013). Modeling mouse, human, and discipline: Epistemic scaffolds in animal behavior genetics. *Social Studies of Science, 43*(1), 3–29.

Ness, S. A. (2013). Ecologies of falling in Yosemite National Park. *Performance Research, 18*(4), 14–21.

Neumark, N., Gibson, R., & van Leeuwen, T. (2010). *Voice: Vocal aesthetics in digital arts and media.* Cambridge: MIT Press.

Nguyen, T. D., Thanh, T. T., Nguyen, L. L., & Huynh, H. T. (2015). On the design of energy efficient environment monitoring station and data collection network based on ubiquitous wireless sensor networks. In *Computing & communication technologies—Research, Innovation, and Vision for the Future (RIVF), 2015 IEEE RIVF international conference* (pp. 163–168). New York, NY: IEEE Communications Society.

Obrist, M. K., Pavan, G., Sueur, J., Riede, K., Llusia, D., & Marquez, R. (2010). Bioacoustics approaches in biodiversity inventories. In J. Eymann, J. Degreef, C. L. Häuser, J. C. Monje, Y. Samyn, & D. Vandan Spiegel (Eds.), *Manual on field recording techniques and protocols for all taxa biodiversity* inventories (pp. 68–99). Brussels: Belgian Development Cooperation.

Olden, J. D. (2006). Biotic homogenization: A new research agenda for conservation biogeography. *Journal of Biogeography, 33*(12), 2027–2039.

Oliveros, P. (2011). Auralizing in the sonosphere: A vocabulary for inner sound and sounding. *Journal of Visual Culture, 10*(2), 162–168.

Pantalony, D. (2009). *Altered sensations: Rudolph Koenig's acoustical workshop in nineteenth-century Paris.* New York: Springer.

Parikka, J. (2010). *Insect media: An archaeology of animals and technology.* Minneapolis: University of Minnesota Press.

Pedwell, C. (2014). *Affective relations: The transnational politics of empathy.* London, UK: Palgrave Macmillan.

Pesci, A., Fabris, M., Conforti, D., Loddo, F., Baldi, P., & Anzidei, M. (2007). Integration of ground-based laser scanner and aerial digital photogrammetry for topographic modelling of Vesuvio volcano. *Journal of Volcanology and Geothermal Research, 162,* 123–138.

Peters, J. D. (2015). *The marvelous clouds: Toward a philosophy of elemental media.* Chicago: University of Chicago Press.

Pieters, B. M., Eindhoven, G. B., Acott, C., & van Zundert, A. A. J. (2015). Pioneers of laryngoscopy: Indirect, direct and video laryngoscopy. *Anaesthesia and Intensive Care, 43*(Suppl. 1), 4–11.

Pijanowski, B. C., Farina, A., Gage, S. H., Dumyahn, S. L., & Krause, B. L. (2011). What is soundscape ecology? An introduction and overview of an emerging new science. *Landscape Ecology, 26*(9), 1213–1232.

Pijanowski, B. C., Villanueva-Rivera, L. J., Dumyahn, S. L., Farina, A., Krause, B. L., Napoletano, B. M., & Pieretti, N. (2011b). Soundscape ecology: The science of sound in the landscape. *BioScience, 61*(3), 203–216.

Pinch, T., & Bijsterveld, K. (Eds.). (2012). *The Oxford handbook of sound studies.* Oxford, UK: Oxford University Press.

Ponomarenko, A., Vincent, O., Pietriga, A., Cochard, H., Badel, É., & Marmottant, P. (2014). Ultrasonic emissions reveal individual cavitation bubbles in water-stressed wood. *Journal of the Royal Society Interface, 11*(99), 20140480.

Portfors, C. V. (2007). Types and functions of ultrasonic vocalizations in laboratory rats and mice. *Journal of the American Association for Laboratory Animal Science, 46*(1), 28–34.

Potts, T. J. (2012). "Dark tourism" and the "kitschification" of 9/11. *Tourist Studies, 12,* 232–249.

Powers, A. (2016). Preserving the quietest places. *The California Sunday Magazine.* Retrieved from https://story.californiasunday.com/quietest-places-on-earth.

Priest, E. (2018). Earworms, daydreams and cognitive capitalism. *Theory, Culture & Society, 35*(1), 141–162.

Prior, N. (2011). Speed, rhythm, and time-space: Museums and cities. *Space and Culture, 14*(2), 197–213.

Radick, G. (2007). *The simian tongue: The long debate about animal language.* Chicago, IL: Chicago University Press.

Rahaim, M. (2019). Object, person, machine, or what. In N. Eidsheim & K. Meizel (Eds.), *The Oxford handbook of voice studies* (pp. 19–34). Oxford: Oxford University Press.

Rancière, J. (2004). Introducing disagreement. *Angelaki, 9*(3), 3–9.

Rempel, R. S., Hobson, K. A., Holborn, G., Wilgenburg, S. L. V., & Elliott, J. (2005). Bioacoustic monitoring of forest songbirds: Interpreter variability and effects of configuration and digital processing methods in the laboratory. *Journal of Field Ornithology, 76*(1), 1–11.

Rickards, L. A. (2015). Metaphor and the anthropocene: Presenting humans as a geological force. *Geographical Research, 5,* 280–287.

Rosen, J. (2017). Sustainability: A greener future. *Nature, 546*(7659), 565–567.

Ross, A. (2012). Song of the earth. In B. Herzogenrath (Ed.), *The farthest place: The music of John Luther Adams* (pp. 13–22). Princeton, NJ: Northeastern University Press.

Rothenberg, D. (2008). *Thousand-mile song: Whale music in a sea of sound.* New York, NY: Basic Books.

Runia, E. (2014). *Moved by the past: Discontinuity and historical mutation.* New York: Columbia University Press.

Ruppaner, A. (1867). The practice of laryngoscopy and rhinoscopy. *New York Medical Journal: A Monthly Record of Medicine and the Collateral Sciences, 6*(1), 1–25.

Rush, J. (1827). *The philosophy of the human voice: Embracing its physiological history; together with a system of principles by which criticism in the art of elocution may be rendered intelligible, and instruction, definite and comprehensive to which is added a brief analysis of song and recitative.* Philadelphia, PA: Printed by J. Maxwell.

Saner, E. (2012). Voice lifts: Something to shout about. *The Guardian.* Retrieved from https://www.theguardian.com/lifeandstyle/2012/sep/23/voice-lift-vocal-cord-treatment.

Sassa, K. (2016). Implementation of the ISDR-ICL Sendai partnerships 2015_2025 for global promotion of understanding and reducing landslide disaster risk. *Landslides, 13,* 211–214.

Sassa, K., & Canuti, P. (Eds.). (2008). *Landslides-disaster risk reduction.* Berlin, Germany: Springer Science & Business Media.

Savage-Rumbaugh, S., Wamba, K., Wamba, P., & Wamba, N. (2007). Welfare of apes in captive environments: Comments on, and by, a specific group of apes. *Journal of Applied Animal Welfare Science, 10*(1), 7–19.

Schafer, R. M. (1993). *The soundscape: Our sonic environment and the tuning of the world.* Rochester, NY: Inner Traditions—Bear & Company.

Schafer, R. M. (1994). *The soundscape.* Rochester, VT: Destiny Books.

Schulze, H. (2018). *The sonic persona: An anthropology of sound.* New York: Bloomsbury Academic.

Scott, T. (2010). *Organization philosophy: Gehlen, Foucault, Deleuze.* London, UK: Palgrave Macmillan.

Seaton, A. V. (1996). Guided by the dark: From Thanatopsis to Thanatourism. *International Journal of Heritage Studies, 2,* 234–244.

Seiler, E. (1884). *The voice in singing.* Philadelphia, PA: J. B. Lippincott & Co.

Servick, K. (2014). Eavesdropping on ecosystems. *Science, 343*(6173), 834–837.

Sha, X. W. (2013). *Poiesis and enchantment in topological matter.* Cambridge: MIT Press.

Shadlen, M. N., & Gold, J. L. (2004). The neurophysiology of decision making as a window on cognition. In M. S. Gazzaniga (Ed.), *The cognitive neurosciences III* (pp. 1229–1243). Cambridge: MIT Press.

Sharpe, C. F. S. (1938). *Landslides and related phenomena: A study of mass-movements of soil and rock.* New York, NY: Pageant Books.

Shepherd, J. (1991). *Music as social text.* London, UK: Polity.

Shepherd, J. (2015). Music, the body, and signifying practices. In K. Devine & J. Shepherd (Eds.), *The Routledge reader on the sociology of music* (pp. 87–96). London, UK: Routledge.

Shields, R. (1991). *Places on the margin: Alternative geographies of modernity*. New York, NY: Routledge.

Shields, R. (2006). Flânerie for cyborgs. *Theory, Culture & Society, 23*(7–8), 209–220.

Shields, R. (2013). *Spatial questions: Cultural topologies and social spatialisation*. London, UK: Sage.

Shields, R. (2017). At the border: Expanding the sociological imagination. Hypothesis of spatial difference and social inequality. *Current Sociology, 65*(4), 533–552.

Sholl, R. (2015). "Stop it, I like it!" Embodiment, masochism, and listening for traumatic pleasure. In S. van Maas (Ed.), *Thresholds of listening: Sound, technics, space* (pp. 153–174). New York, NY: Fordham University Press.

Sicular, S. (2013, January 22). *Big Data is falling into the trough of disillusionment*. Retrieved from Gartner database. https://blogs.gartner.com/svetlana-sicular/big-data-is-falling-into-the-trough-of-disillusionment/.

Simpson, P. (2012). Apprehending everyday rhythms: Rhythmanalysis, time-lapse photography, and the space-times of street performance. *Cultural Geographies, 19*(4), 423–445.

Slabbekoorn, H., & Peet, M. (2003). Birds sing at a higher pitch in urban noise. *Nature, 424*, 267.

Smith, D. N. (2012). *Sounding/silence: Martin Heidegger and the limits of poetics*. New York: Fordham University Press.

Smith, H., & Dean, R. T. (2003). Voicescapes and sonic structures in the creation of sound technodrama. *Performance Research, 8*(1), 112–123.

Smith, K. (2013). *Environmental hazards: Assessing risk and reducing disaster*. Oxford, UK: Routledge.

Spencer, H. (1891). The origin and function of music. In *Essays: Scientific, political, and speculative, 2*. London: Williams and Norgate.

Spencer, H. (2015). The origin and function of music. In J. Shepherd & K. Devine (Eds.), *The Routledge reader on the sociology of music* (pp. 43–50). London: Routledge.

Stähli, M., Sättele, M., Huggel, C., McArdell, B. W., Lehmann, P., Van Herwijnen, A., ... Springman, S. M. (2015). Monitoring and prediction in early warning systems for rapid mass movements. *Natural Hazards and Earth System Sciences, 15*, 905–917.

Stahr, A., & Langenscheidt, E. (2014). *Landforms of high mountains*. Berlin, Germany: Springer.

Sterne, J. (2003). *The audible past: Cultural origins of sound reproduction*. Durham, NC: Duke University Press.

Sterne, J. (Ed.). (2012). *The sound studies reader.* London, UK: Routledge.

Sterne, J., & Akiyama, M. (2012). The recording that never wanted to be heard and other stories of sonification. In T. Pinch & K. Bijsterveld (Eds.), *The Oxford handbook of sound studies* (pp. 544–560). Oxford: Oxford University Press.

Stiegler, B. (1998). *Technics & time 1: The fault of epimetheus* (R. Beardsworth, Trans.). Stanford, USA: Stanford University Press.

Stiegler, B. (2004). *De la misère symbolique 1.* Paris, France: Galilée.

Stiegler, B. (2010). *Taking care of youth and the generations.* Stanford, CA: Stanford University Press.

Stowell, D., Benetos, E., & Gill, L. F. (2017). On-bird sound recordings: Automatic acoustic recognition of activities and contexts. *IEEE/ACM Transactions on Audio, Speech, and Language Processing, 25*(6), 1193–1206.

Stowell, D., Giannoulis, D., Benetos, E., Lagrange, M., & Plumbey, M. D. (2015). Detection and classification of acoustic scenes and events. *IEEE Transactions on Multimedia, 17*(10), 1733–1746.

Sueur, J., & Farina, A. (2015). Ecoacoustics: The ecological investigation and interpretation of environmental sound. *Biosemiotics, 8*(3), 493–502.

Supper, A. (2014). Sublime frequencies: The construction of sublime listening experiences in the sonification of scientific data. *Social Studies of Science, 44*(1), 34–58.

Szostak, R., Gnoli, C., & López-Huertas, M. (2016). *Interdisciplinary knowledge organization.* New York: Springer International Publishing.

Tarchi, D., Casagli, N., Moretti, S., Leva, D., & Sieber, A. J. (2003). Monitoring landslide displacements by using ground-based synthetic aperture radar interferometry: Application to the Ruinon landslide in the Italian Alps. *Journal of Geophysical Research: Solid Earth, 108*(B8). Retrieved from https://doi.org/10.1029/2002JB002204.

Tegeler, A. K., Morrison, M. L., & Szewczak, J. M. (2012). Using extended-duration audio recordings to survey avian species. *Wildlife Society Bulletin, 36*(1), 21–29.

Terzaghi, K. (1950). Mechanism of landslide. In S. Paige (Ed.), *Application of geology to engineering practice* (pp. 82–123). New York, NY: Geological Society of America.

Terzaghi, K. (1961). Stability of steep slopes on hard unweathered rock. *Geotechnique, 12,* 251–270.

Thomaidis, K., & Macpherson, B. (Eds.). (2015). *Voice studies: Critical approaches to process, performance and experience.* London, UK: Routledge.

Thoreau, D. (1885). *The writings of Henry David Thoreau* (Vol. 6). Boston: Houghton Mifflin.

Tiainen, M. (2013). Revisiting the voice in media and as medium: New materialist propositions. *NECSUS: European Journal of Media Studies, 2*(2), 383–406.

Tønnessen, M. (2015). The biosemiotic glossary project: Agent, agency. *Biosemiotics, 8,* 125–143.

Torino, L. (2015). You can actually hear the climate changing. *Outside.* Retrieved from https://www.outsideonline.com/2035701/you-can-actually-hear-climate-changing.

Towsey, M., Wimmer, J., Williamson, I., & Roe, P. (2014). The use of acoustic indices to determine avian species richness in audio-recordings of the environment. *Ecological Informatics, 21,* 110–119.

Truax, B. (2001). *Acoustic communication* (Vol. 1). Westport: Greenwood Publishing Group.

Upton, B. (2015). *The aesthetic of play.* Cambridge: MIT Press.

Urry, J. (2003). *Global complexity.* Cambridge, UK: Polity.

Urry, J., & Larsen, J. (2011). *The tourist gaze 3.0.* London, UK: Sage.

Vallee, M. (2017). The rhythm of echoes and echoes of violence. *Theory, Culture & Society, 34*(1), 97–114.

Vallee, M. (2018, September 17). *Tour de Frank.* Field notes at the Frank Slide site.

Vartan, S. (2016). We're changing the way the world sounds: Noise impacts ecosystems in more ways than you might think. *Mother Nature Network.* Retrieved from http://www.mnn.com/earth-matters/wilderness-resources/blogs/we-are-changing-way-world-sounds.

Viel, J. M. (2014). *Habitat preferences of the common nighthawk (Chordeiles minor) in cities and villages in southeastern Wisconsin* (Doctoral dissertation). Retrieved from https://dc.uwm.edu/etd/516/). The University of Wisconsin-Milwaukee.

Virno, P. (2015). *When the word becomes flesh: Language and human nature.* Los Angeles: Semiotext(e).

von Uexküll, J. (1982). Glossary. *Semiotica, 42*(1), 83–87.

von Uexküll, J. J. (1934/2010). *A foray into the worlds of animals and humans: With a theory of meaning* (J. D. O'Neil, Trans.). Minneapolis: University of Minnesota Press.

Wachinger, G., Renn, O., Begg, C., & Kuhlicke, C. (2013). The risk perception paradox—Implications for governance and communication of natural hazards. *Risk Analysis, 33,* 1049–1065.

Wallheimer, B. (2011, March 23). New scientific study will study ecological importance of sounds. *Science News*. Retrieved from https://www.sciencedaily.com/releases/2011/03/110301122154.htm.

Wambacq, J., Ross, D., & Buseyne, B. (2016). "We have to become the quasi-cause of nothing—of nihil": An interview with Bernard Stiegler. *Theory, Culture & Society* (Online First). Retrieved from https://doi.org/10.1177/0263276416651932.

Watson, M. C. (2016). On multispecies mythology: A critique of animal anthropology. *Theory, Culture & Society, 33*(5), 159–172.

Wells, W. A. (1946). Benjamin Guy Babington: Inventor of the laryngoscope. *The Laryngoscope, 56*(8), 443–454.

Westerkamp, H. (2001). Speaking from inside the soundscape. In D. Rothenberg & M. Ulvaeus (Eds.), *The book of music and nature: An anthology of sounds, words, thoughts.* Middletown, CT: Wesleyan University Press.

Wilson, D. (2005). *Triumph and tragedy in the Crowsnest Pass.* Surrey, UK: Heritage House.

Windsor, T. (1863, January–April). On the discovery of the laryngoscope. *The British and Foreign Medicochirurgical Review, Or, Quarterly Journal of Practical Medicine and Surgery, 31*, 209–210.

Young, M. (2015). *Singing the body electric: The human voice and sound technology.* London, UK: Ashgate.

Younis, R. T., & Lazar, R. H. (2002). History and current practice of tonsillectomy. *The Laryngoscope, 112*(S100), 3–5.

Zak, A. (2001). *The poetics of rock: Cutting tracks, making records.* Berkeley: University of California Press.

Index

The manufacturer's authorised representative in the EU is Springer
Nature Customer Service Centre GmbH, Europaplatz 3, 69115 Heidelberg,
Germany. If you have any concerns regarding our products, please
contact ProductSafety@springernature.com

Printed and bound by CPI Group (UK) Ltd, Croydon, CR0 4YY
29/04/2026
02099478-0007